TEXTILE FIBERS, DYES, FINISHES, AND PROCESSES

TEXTILE SERIES

Howard L. Needles, Editor

FABRIC FORMING SYSTEMS
By *Peter Schwartz, Trever Rhodes* and *Mansour Mohamed*

TEXTILE IDENTIFICATION, CONSERVATION, AND PRESERVATION
By *Rosalie Rosso King*

TEXTILE MARKETING MANAGEMENT
By *Gordon A. Berkstresser III*

TEXTILE WET PROCESSES: Vol. I. Preparation of Fibers and Fabrics
By *Edward S. Olson*

TEXTILE FIBERS, DYES, FINISHES, AND PROCESSES: A Concise Guide
By *Howard L. Needles*

AUTOMATION AND ROBOTICS IN THE TEXTILE AND APPAREL INDUSTRIES
Edited by *Gordon A. Berkstresser III* and *David R. Buchanan*

Other Title

TEXTILE DYEING OPERATIONS: Chemistry, Equipment, Procedures, and Environmental Aspects
By *S.V. Kulkarni, C.D. Blackwell, A.L. Blackard, C.W. Stackhouse* and *M.W. Alexander*

TEXTILE FIBERS, DYES, FINISHES, AND PROCESSES

A Concise Guide

by

Howard L. Needles

University of California, Davis
Davis, California

np | **NOYES PUBLICATIONS**
Park Ridge, New Jersey, U.S.A.

Published in the United States of America by
Noyes Publications
Mill Road, Park Ridge, New Jersey 07656

Transferred to Digital Printing, 2011

Printed and bound in the United Kingdom

Library of Congress Cataloging-in-Publication Data

Needles, Howard L.
Textile fibers, dyes, finishes, and processes.

Bibliography: p.
Includes index.
1. Textile fibers. 2. Textile finishing. 3. Dyes
and dyeing--Textile fibers. I. Title.
TS1540.N434 1986   677'.028   86-5203
ISBN 0-8155-1076-4

Preface

Fibers from natural sources have been used for thousands of years for producing textiles and related products. With the advent of the spinning jet in the mid-19th century, fibers could be formed by forcing dissolved polymeric materials through a small orifice (spinneret) into a coagulating bath. Regenerated natural and synthetic man-made fibers have been formed by this basic spinning technique or variations thereof since then. By the turn of the 20th century, rayon, a regenerated cellulosic and the first man-made fiber of commercial importance, was in full production. By the 1920s the cellulose derivatives acetate and triacetate were introduced as fibers of commerce, and inorganic glass fibers appeared during the mid-1930s. The first synthetic fiber (nylon) chemically synthesized from basic monomeric units and based on petroleum feedstocks appeared in the late 1930s. The advent of nylon marked a new era for fiber production, and several new types of synthetic fibers, including polyester, acrylic, modacrylic, polyolefin, and vinyl fibers, appeared in the 1940s, 1950s, and 1960s.

In less than 40 years we have gone from a period where fibers were available only from natural or regenerated sources to a time where a broad spectrum of fibers are available. The wide range of properties available in fibers today has greatly expanded the applications and areas in which fibers can be used. Even with such a range of properties available in fibers, each class of fiber has inherent deficiencies that require that chemical finishes or physical modifications be applied to the fiber. Also, addition of color to the fiber through dyeing or printing is necessary to meet the demand of the consumer for a wide spectrum of colors and patterns in textile products. Since 1945 a number of new textile processes have been introduced providing unique methods to form yarns and textile substrates of widely varying structure and properties. This book addresses itself to the structure and properties of textile fibers, dyes, and finishes and the processes used in fiber, yarn, and substrate formation and in dyeing and finishing of these substrates.

Owing to the growing number, types, and complexity of fibers now available for use in consumer textiles, students or professionals in textiles, textiles and clothing, and textile science need not only a listing of fibers and fiber properties but also a firm foundation in the relationship of fiber structure to the physical and chemical properties of fibers, as well as the consumer end-use properties that result in textiles made from these fibers. They also need to be acquainted with the processes used in formation of textile fibers, yarns, and fabric substrates and in dyeing and finishing these substrates. Textbooks in consumer textiles often stress the more aesthetic areas of textiles, whereas textbooks in textile chemistry and textile physics present a highly rigorous approach to the field. A book which lies between these two extremes would be of value to those with an intermediate understanding of the physical sciences. Thus this book discusses textile fibers, dyes, finishes, and processes using this intermediate approach, presenting in a concise manner the underlying principles of textile chemistry, physics, and technology. It should be an aid to students and professionals in textiles, textiles and clothing, and textile science, who desire a basic knowledge of textile fibers, finishes, and processes and their related consumer end-use. The book should also serve as a sourcebook of information within the textile and apparel industries.

I thank my colleagues and students who have contributed in numerous ways to this book. I especially thank Barbara Brandon for her expert preparation of the book for print.

University of California, Davis   Howard L. Needles
March, 1986

ABOUT THE AUTHOR

Howard L. Needles is presently Professor of Textiles and Materials Science at the University of California, Davis. After receiving his doctorate in organic chemistry from the University of Missouri in 1963, he began his career conducting research on wool and related model systems. His research was then extended to include synthetic fibers and the effect of chemical modification on the dyeing and color properties of these fibers. He has also continued his studies at North Carolina State University and at the University of Leeds, England, and is also Program Chairman of the Cellulose, Paper and Textile Division of the American Chemical Society.

Contents

V. TEXTILE MAINTENANCE

I. Fiber Theory, Formation, and Characterization

1. Fiber Theory and Formation

INTRODUCTION

The word "textile" was originally used to define a woven fabric and the processes involved in weaving. Over the years the term has taken on broad connotations, including the following: (1) staple filaments and fibers for use in yarns or preparation of woven, knitted, tufted or non-woven fabrics, (2) yarns made from natural or man-made fibers, (3) fabrics and other products made from fibers or from yarns, and (4) apparel or other articles fabricated from the above which retain the flexibility and drape of the original fabrics. This broad definition will generally cover all of the products produced by the textile industry intended for intermediate structures or final products.

Textile fabrics are planar structures produced by interlacing or entangling yarns or fibers in some manner. In turn, textile yarns are continuous strands made up of textile fibers, the basic physical structures or elements which makes up textile products. Each individual fiber is made up of millions of individual long molecular chains of discrete chemical structure. The arrangement and orientation of these molecules within the individual fiber, as well as the gross cross section and shape of the fiber (morphology), will affect fiber properties, but by far the molecular structure of the long molecular chains which make up the fiber will determine its basic physical and chemical nature. Usually, the polymeric molecular chains found in fibers have a definite chemical sequence which repeats itself along the length of the molecule. The total number of units which repeat themselves in a chain (n) varies from a few units to several hundred and is referred to as the degree of polymerization (DP) for molecules within that fiber.

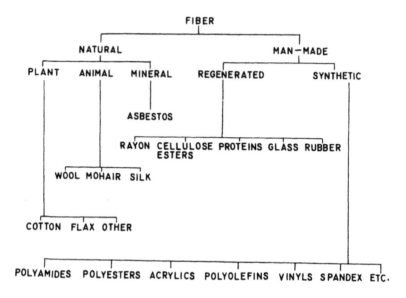

Figure 1-1. Classification of natural and man-made fibers.

FIBER CLASSIFICATION

Textile fibers are normally broken down into two main classes, natural and man-made fibers. All fibers which come from natural sources (animals, plants, etc.) and do not require fiber formation or reformation are classed as natural fibers. Natural fibers include the protein fibers such as wool and silk, the cellulose fibers such as cotton and linen, and the mineral fiber asbestos. Man-made fibers are fibers in which either the basic chemical units have been formed by chemical synthesis followed by fiber forma- tion or the polymers from natural sources have been dissolved and regener- ated after passage through a spinneret to form fibers. Those fibers made by chemical synthesis are often called synthetic fibers, while fibers re- generated from natural polymer sources are called regenerated fibers or natural polymer fibers. In other words, all synthetic fibers and regenerated fibers are man-made fibers, since man is involved in the actual fiber formation process. In contrast, fibers from natural sources are provided by nature in ready-made form.

The synthetic man-made fibers include the polyamides (nylon), polyesters, acrylics, polyolefins, vinyls, and elastomeric fibers, while the regenerated fibers include rayon, the cellulose acetates, the regenerated proteins, glass and rubber fibers. Figure 1-1 shows a classification chart for the major fibers.

Another method of classifying fibers would be according to chemical structure without regard of the origin of the fiber and its starting materials. In this manner all fibers of similar chemical structure would be classed together. The natural man-made fiber classification given in Figure 1-1 does this to a certain extent. In this way, all fibers having the basic cellulosic unit in their structures would be grouped together rather than separated into natural and man-made fibers. This book essentially presents the fibers in groups of similar basic chemical structure, with two exceptions. In one case the elastomeric fibers have been grouped together due to their exceptional physical property, high extensibility and recovery. In the other case, new fibers which do not properly "fit" into any one category have been placed in a separate chapter. An outline for the arrangement for fibers by chemical class as presented in this sourcebook follows:

Cellulosic Fibers
    Cotton
    Flax
    Other natural cellulosic
       fibers
    Rayon

Cellulosic Ester Fibers
    Acetate
    Triacetate

Protein (Natural Polyamide)
 Fibers
    Wool
    Silk
    Other natural and regen-
      erated protein fibers

Polyamide (Nylon) Fibers
    Nylon 6 and 6,6
    Aramid
    Other nylon fibers

Polyester Fibers
    Polyethylene terephthalate
    Poly-1,4-cyclohexylenedi-
      methylene terephthalate
    Other polyester fibers

Acrylic and Modacrylic Fibers
    Acrylic
    Modacrylic
    Other acrylics

Polyolefin Fibers
    Polyethylene
    Polypropylene

Vinyl Fibers
    Vinyon
    Vinal
    Vinyon-vinal matrix
    Saran
    Polytetrafluoroethylene

Elastomeric Fibers
    Rubber
    Spandex
    Other elastomeric fibers

Mineral and Metallic Fibers
    Glass
    Inorganic
    Asbestos
    Metallic

Miscellaneous Fibers
    Novaloid
    Carbon
    Poly(m-phenylenediben-
      zimidazole)
    Polyimide

FIBER PROPERTIES

There are several primary properties necessary for a polymeric material to make an adequate fiber: (1) fiber length to width ratio, (2) fiber uniformity, (3) fiber strength and flexibility, (4) fiber extensibility and elasticity, and (5) fiber cohesiveness.

Certain other fiber properties increase its value and desirability in its intended end-use but are not necessary properties essential to make a fiber. Such secondary properties include moisture absorption characteristics, fiber resiliency, abrasion resistance, density, luster, chemical resistance, thermal characteristics, and flammability. A more detailed description of both primary and secondary properties follows.

Primary Properties

Fiber Length to Width Ratio: Fibrous materials must have sufficient length so that they can be made into twisted yarns. In addition, the width of the fiber (the diameter of the cross section) must be much less than the overall length of the fiber, and usually the fiber diameter should be 1/100 of the length of the fiber. The fiber may be "infinitely" long, as found with continuous filament fibers, or as short as 0.5 inches (1.3 cm), as found in staple fibers. Most natural fibers are staple fibers, whereas man-made fibers come in either staple or filament form depending on processing prior to yarn formation.

Fiber Uniformity: Fibers suitable for processing into yarns and fabrics must be fairly uniform in shape and size. Without sufficient uniformity of dimensions and properties in a given set of fibers to be twisted into yarn, the actual formation of the yarn may be impossible or the resulting yarn may be weak, rough, and irregular in size and shape and unsuitable for textile usage. Natural fibers must be sorted and graded to assure fiber uniformity, whereas synthetic fibers may be "tailored" by cutting into appropriate uniform lengths to give a proper degree of fiber uniformity.

Fiber Strength and Flexibility: A fiber or yarn made from the fiber must possess sufficient strength to be processed into a textile fabric or other textile article. Following fabrication into a textile article, the resulting textile must have sufficient strength to provide adequate durability during end-use. Many experts consider a single fiber strength of 5 grams per denier to be necessary for a fiber suitable in most textile applications, although certain fibers with strengths as low as 1.0 gram per denier have been found suitable for some applications.

The strength of a single fiber is called the tenacity, defined as the force per unit linear density necessary to break a known unit of that fiber. The breaking tenacity of a fiber may be expressed in grams per denier (g/d) or grams per tex (g/tex). Both denier and tex are units of linear density (mass per unit of fiber length) and are defined as the number of grams of fiber measuring 9000 meters and 1000 meters, respectively. As a result, the denier of a fiber or yarn will always be 9 times the tex of the same fiber. Since tenacities of fibers or yarns are obtained by dividing the force by denier or tex, the tenacity of a fiber in grams per denier will be 1/9 that of the fiber tenacity in grams per tex.

As a result of the adaption of the International System of Units, referred to as SI, the appropriate length unit for breaking tenacity becomes kilometer (km) of breaking length or Newtons per tex (N/tex) and will be equivalent in value to g/tex.

The strength of a fiber, yarn, or fabric can be expressed in terms of force per unit area, and when expressed in this way the term is tensile strength. The most common unit used in the past for tensile strength has been pounds force per square inch or grams force per square centimeter. In SI units, the pounds force per square inch x 6.895 will become kilopascals (kPa) and grams force per square centimeter x 9.807 will become megapascals (MPa).

A fiber must be sufficiently flexible to go through repeated bending without significant strength deterioration or breakage of the fiber. Without adequate flexibility, it would be impossible to convert fibers into yarns and fabrics, since flexing and bending of the individual fibers is a necessary part of this conversion. In addition, individual fibers in a textile will be subjected to considerable bending and flexing during end-use.

Fiber Extensibility and Elasticity: An individual fiber must be able to undergo slight extensions in length (less than 5%) without breakage of the fiber. At the same time the fiber must be able to almost completely recover following slight fiber deformation. In other words, the extension deformation of the fiber must be nearly elastic. These properties are important because the individual fibers in textiles are often subjected to sudden stresses, and the textile must be able to give and recover without significant overall deformation of the textile.

Fiber Cohesiveness: Fibers must be capable of adhering to one another when spun into a yarn. The cohesiveness of the fiber may be due to the shape and contour of the individual fibers or the nature of the surface of the fibers. In addition, long-filament fibers by virtue of their length can be twisted together to give stability without true cohesion between fibers. Often the term "spinning quality" is used to state the overall attractiveness of fibers for one another.

Secondary Properties

Moisture Absorption and Desorption: Most fibers tend to absorb moisture (water vapor) when in contact with the atmosphere. The amount of water absorbed by the textile fiber will depend on the chemical and physical structure and properties of the fiber, as well as the temperature and humidity of the surroundings. The percentage absorption of water vapor by a fiber is often expressed as its moisture regain. The regain is determined by weighing a dry fiber, then placing it in a room set to standard temperature and humidity (21° ± 1° C and 65% relative humidity [RH] are commonly used). From these measurements, the percentage moisture regain of the fiber is determined:

$$\text{Percentage regain} = \frac{\text{Conditioned weight - Dry weight}}{\text{Dry weight}} \times 100\%$$

Percentage moisture content of a fiber is the percentage of the total weight of the fiber which is due to the moisture present, and is obtained from the following formula:

$$\text{Percentage moisture content} = \frac{\text{Conditioned weight - Dry weight}}{\text{Conditioned weight}} \times 100\%$$

The percentage moisture content will always be the smaller of the two values.

Fibers vary greatly in their regain, with hydrophobic (water-repelling) fibers having regains near zero and hydrophilic (water-seeking) fibers like cotton, rayon, and wool having regains as high as 15% at 21°C and 65% RH. The ability of fibers to absorb high regains of water affects the basic properties of the fiber in end-use. Absorbent fibers are able to absorb large amounts of water before they feel wet, an important factor where absorption of perspiration is necessary. Fibers with high regains will be easier to process, finish, and dye in aqueous solutions, but will dry more slowly. The low regain found for many man-made fibers makes them quick drying, a distinct advantage in certain applications. Fibers with high regains are often desirable because they provide a "breathable" fabric which can conduct moisture from the body to the outside atmosphere readily, due to their favorable moisture absorption-desorption properties. The tensile properties of fibers as well as their dimensional properties are known to be affected by moisture.

Fiber Resiliency and Abrasion Resistance: The ability of a fiber to absorb shock and recover from deformation and to be generally resistant to abrasion forces is important to its end-use and wear characteristics. In consumer use, fibers in fabrics are often placed under stress through compression, bending, and twisting (torsion) forces under a variety of temperature and humidity conditions. If the fibers within the fabric possess good elastic recovery properties from such deformative actions, the fiber has good resiliency and better overall appearance in end-use. For example, cotton and wool show poor wrinkle recovery under hot moist conditions, whereas polyester exhibits good recovery from deformation as a result of its high resiliency. Resistance of a fiber to damage when mobile forces or stresses come in contact with fiber structures is referred to as abrasion resistance. If a fiber is able to effectively absorb and dissipate these forces without damage, the fiber will show good abrasion resistance. The toughness and hardness of the fiber is related to its chemical and physical structure and morphology of the fiber and will influence the abrasion of the fiber. A rigid, brittle fiber such as glass, which is unable to dissipate the forces of abrasive action, results in fiber damage and breakage, whereas a tough but more plastic fiber such as polyester shows better resistance to abrasion forces. Finishes can affect fiber properties including resiliency and abrasion resistance.

Luster: Luster refers to the degree of light that is reflected from the surface of a fiber or the degree of gloss or sheen that the fiber possesses. The inherent chemical and physical structure and shape of the fiber can affect the relative luster of the fiber. With natural fibers the luster of the fiber is dependent on the morphological form that nature gives the fiber, although the relative luster can be changed by chemical and/or physical treatment of the fiber as found in processes such as mercerization of cotton. Man-made fibers can vary in luster from bright to dull depending on the amount of delusterant added to the fiber. Delusterants such as titanium dioxide tend to scatter and absorb light, thereby making the fiber appear duller. The desirability of luster for a given fiber application will vary and is often dependent on the intended end-use of the fiber in a fabric or garment form and on current fashion trends.

Resistance to Chemicals in the Environment: A textile fiber to be useful must have reasonable resistance to chemicals it comes in contact with in its environment during use and maintenance. It should have resistance to oxidation by oxygen and other gases in the air, particularly in the presence of light, and be resistant to attack by microorganisms and other biological agents. Many fibers undergo light-induced reactions, and fibers from natural sources are susceptible to biological attack, but such deficiencies can be minimized by treatment with appropriate finishes. Textile fibers come in contact with a large range of chemical agents on laundering and dry cleaning and must be resistant from attack under such conditions.

Density: The density of a fiber is related to its inherent chemical structure and the packing of the molecular chains within that structure. The density of a fiber will have a noticeable effect on its aesthetic appeal and its usefulness in given applications. For example, glass and silk fabrics of the same denier would have noticeable differences in weight due to their broad differences in density. Fishnets of polypropylene fibers are of great utility because their density is less than that of water. Densities are usually expressed in units of grams per cubic centimeter, but in SI units will be expressed as kilograms per cubic meter, which gives a value 1000 times larger.

Thermal and Flammability Characteristics: Fibers used in textiles must be resistant to wet and dry heat, must not ignite readily when coming in contact with a flame, and ideally should self-extinguish when the flame is removed. Heat stability is particularly important to a fiber during dyeing and finishing of the textile and during cleaning and general maintenance by the consumer. Textile fibers for the most part are made up of organic polymeric materials containing carbon and burn on ignition from a flame or other propagating source. The chemical structure of a fiber establishes its overall flammability characteristics, and appropriate textile finishes can reduce the degree of flammability. A number of Federal, state, and local statutes eliminate the most dangerous flammable fabrics from the marketplace.

Primary Fiber Properties from an Engineering Perspective

The textile and polymer engineer must consider a number of criteria essential for formation, fabrication, and assembly of fibers into textile substrates. Often the criteria used will be similar to those set forth above concerning end-use properties. Ideally a textile fiber should have the following properties:

1. A melting and/or decomposition point above 220°C.
2. A tensile strength of 5 g/denier or greater.
3. Elongation at break above 10% with reversible elongation up to 5% strain.
4. A moisture absorptivity of 2%-5% moisture uptake.
5. Combined moisture regain and air entrapment capability.
6. High abrasion resistance.
7. Resistance to attack, swelling, or solution in solvents, acids, and bases.
8. Self-extinguishing when removed from a flame.

FIBER FORMATION AND MORPHOLOGY

Fiber morphology refers to the form and structure of a fiber, including the molecular arrangement of individual molecules and groups of molecules within the fiber. Most fibers are organic materials derived from carbon combined with other atoms such as oxygen, nitrogen, and halogens. The basic building blocks that organic materials form as covalently-bonded organic compounds are called monomers. Covalent bonds involve the sharing of electrons between adjacent atoms within the monomer, and the structure trons between adjacent atoms within the monomer, and the structure of the monomer is determined by the type, location, and nature of bonding of atoms within the monomer and by the nature of covalent bonding between atoms. Monomers react or condense to form long-chain molecules called polymers made up of a given number (n) of monomer units which are the basic building unit of fibers. On formation into fibers and orientation by natural or mechanical means the polymeric molecules possess ordered crystalline and nonordered amorphous areas, depending on the nature of the polymer and the relative packing of molecules within the fiber. For a monomer A the sequence of events to fiber formation and orientation would appear as shown in Figure 1-2.

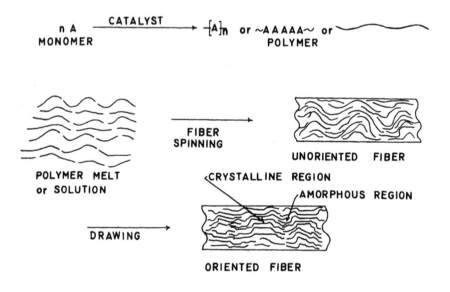

Figure 1-2. Polymerization sequence and fiber formation.

Polymers with repeating units of the same monomer (A_n) would be referred to as homopolymers. If a second unit B is introduced into the basic structure, structures which are copolymers are formed with structures as outlined in Figure 1-3.

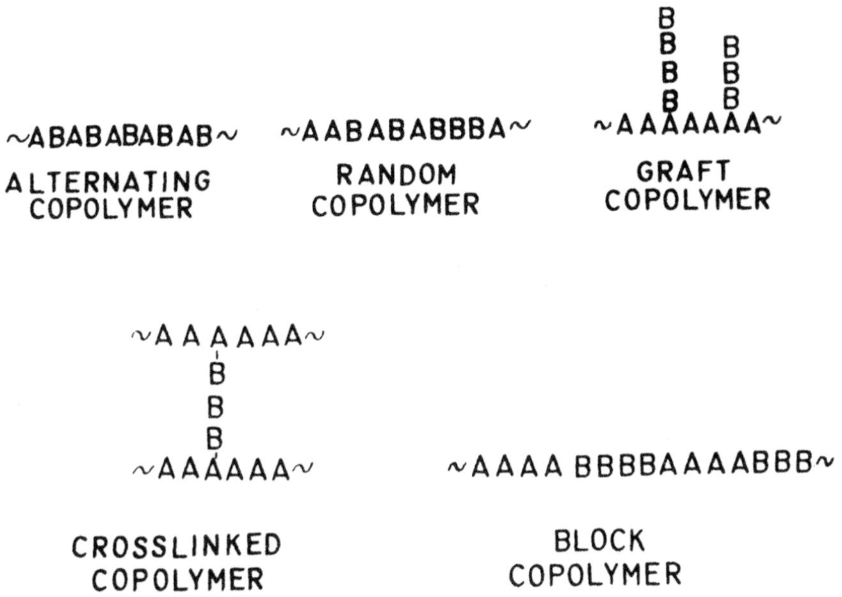

~ABABABABAB~

ALTERNATING
COPOLYMER

~AABABABBBA~

RANDOM
COPOLYMER

```
              B
              B
              B     B
              B     B
              B     B
~AAAAAAA~
```

GRAFT
COPOLYMER

```
~A A A A AA~
      B
      B
      B
~AAAAAA~
```

CROSSLINKED
COPOLYMER

~AAAA BBBBAAAABBB~

BLOCK
COPOLYMER

Figure 1-3. Copolymer structures.

Polymer Formation

Synthetic polymers used to form fibers are often classified on the basis of their mechanism of polymerization—step growth (condensation) or chain growth (addition) polymerization. Step growth polymerization involves multifunctional monomers which undergo successive condensation with a second monomer or with itself to form a dimer, which in turn condenses with another dimer to form a tetramer, etc., usually with loss of a small molecule such as water. Chain growth involves the instantaneous growth of a long molecular chain from unsaturated monomer units, followed by initiation of a second chain, etc. The two methods are outlined below schematically:

Step growth: $nA \rightarrow \frac{n}{2} AA \rightarrow \frac{n}{4} AAAA \rightarrow \ldots$

Chain growth: $nA \rightarrow (A)_n$ $nA \rightarrow (A)_n$ $nA \rightarrow (A)_n$

The average number of monomer repeating units in a polymer chain (n) is often referred to also as the degree of polymerization, DP. The DP must exceed an average 20 units in most cases to give a polymer of sufficient molecular size to have desirable fiber-forming properties. The overall breadth of distribution of molecular chain lengths in the polymer will affect the ultimate properties of the fibers, with wide polymer size distributions leading to an overall reduction of fiber properties. Although the polymers from natural fibers and regenerated natural fibers do not undergo polymerization by the mechanisms found for synthetic fibers, most natural polymers have characteristic repeating units and high degrees of polymerization and are related to step growth polymers. Basic polymeric structures for the major fibers are given in Figure 1-4.

Fiber Spinning

Although natural fibers come in a morphological form determined by nature, regenerated and synthetic man-made fibers can be "tailor-made" depending on the shape and dimensions of the orifice (spinning jet) that the polymer is forced through to form the fiber. There are several methods used to spin a fiber from its polymer, including melt, dry, wet, emulsion, and suspension spinning.

Melt spinning is the least complex of the methods. The polymer from which the fiber is made is melted and then forced through a spinneret and into air to cause solidification and fiber formation.

Dry and wet spinning processes involve dissolving the fiber-forming polymer in an appropriate solvent, followed by passing a concentrated solution (20%-50% polymer) through the spinneret and into dry air to evaporate the solvent in the case of dry spinning or into a coagulating bath to cause precipitation or regeneration of the polymer in fiber form in the case of wet spinning. There is a net contraction of the spinning solution on loss of solvent. If a skin of polymer is formed on the fiber followed by diffusion of the remainder of the solvent from the core of the forming fiber, the cross section of the fiber as it contracts may collapse to form an irregular popcornlike cross section.

Emulsion spinning is used only for those fiber-forming polymers that are insoluble. Polymer is mixed with a surface-active agent (detergent) and possibly a solvent and then mixed at high speeds with water to form an emulsion of the polymer. The polymer is passed through the spinneret and into a coagulating bath to form the fiber.

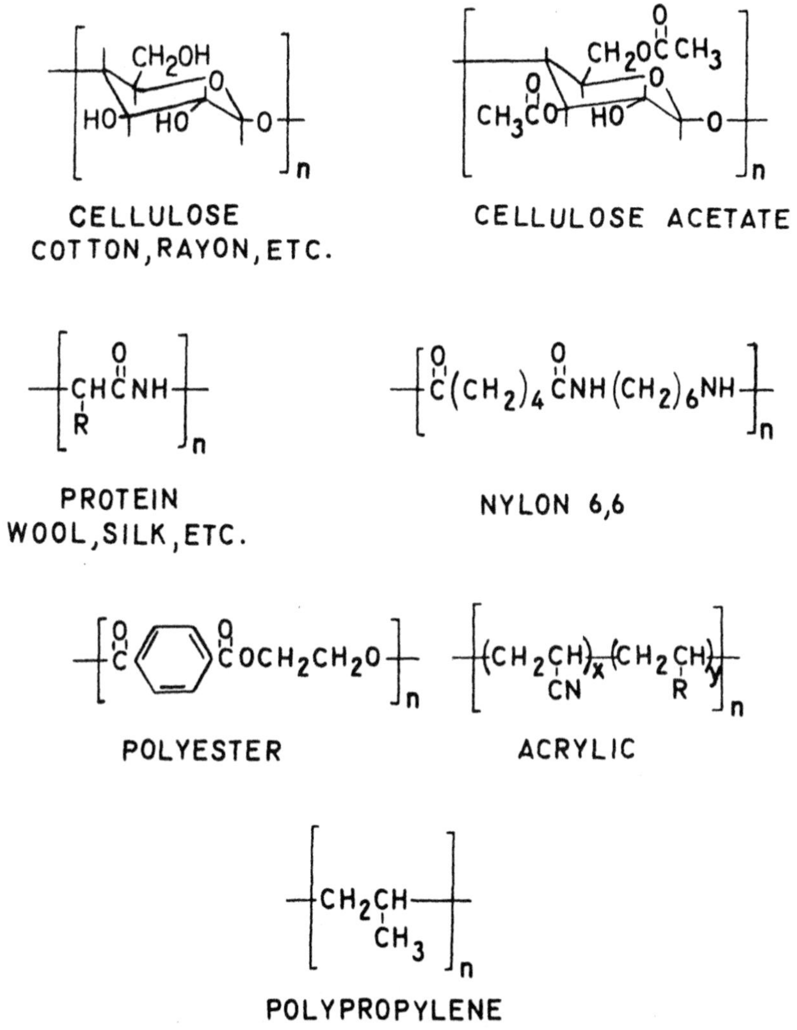

CELLULOSE
COTTON, RAYON, ETC.

CELLULOSE ACETATE

PROTEIN
WOOL, SILK, ETC.

NYLON 6,6

POLYESTER

ACRYLIC

POLYPROPYLENE

Figure 1-4. Basic polymeric structures for major fibers.

In suspension spinning, the polymer is swollen and suspended in a swelling solvent. The swollen suspended polymer is forced through the spinneret into dry hot air to drive off solvent or into a wet non-solvent bath to cause the fiber to form through coagulation.

The spinning process can be divided into three steps:

(1) flow of spinning fluid within and through the spinneret under high stress and sheer.

(2) exit of fluid from the spinneret with relief of stress and an increase in volume (ballooning of flow).

(3) elongation of the fluid jet as it is subjected to tensile force as it cools and solidifies with orientation of molecular structure within the fiber.

Common cross sections of man-made fibers include round, trilobal, pentalobal, dog-bone, and crescent shapes. When two polymers are used in fiber formation as in bicomponent or biconstituent fibers, the two components can be arranged in a matrix, side-by-side, or sheath-core configuration. Round cross sections are also found where skin formation has caused fiber contraction and puckering (as with rayons) has occurred or where the spinneret shape has provided a hollow fiber. Complex fiber cross-sectional shapes with special properties are also used. See Figure 1-5.

Fiber Drawing and Morphology

On drawing and orientation the man-made fibers become smaller in diameter and more crystalline, and imperfections in the fiber morphology are improved somewhat. Side-by-side bicomponent or biconstituent fibers on drawing become wavy and bulky.

In natural fibers the orientation of the molecules within the fiber is determined by the biological source during the growth and maturity process of the fiber.

The form and structure of polymer molecules with relation to each other within the fiber will depend on the relative alignment of the molecules in relationship to one another. Those areas where the polymer chains are closely aligned and packed close together are crystalline areas within the fiber, whereas those areas where there is essentially no molecular alignment are referred to as amorphous areas. Dyes and finishes can penetrate the amorphous portion of the fiber, but not the ordered crystalline portion.

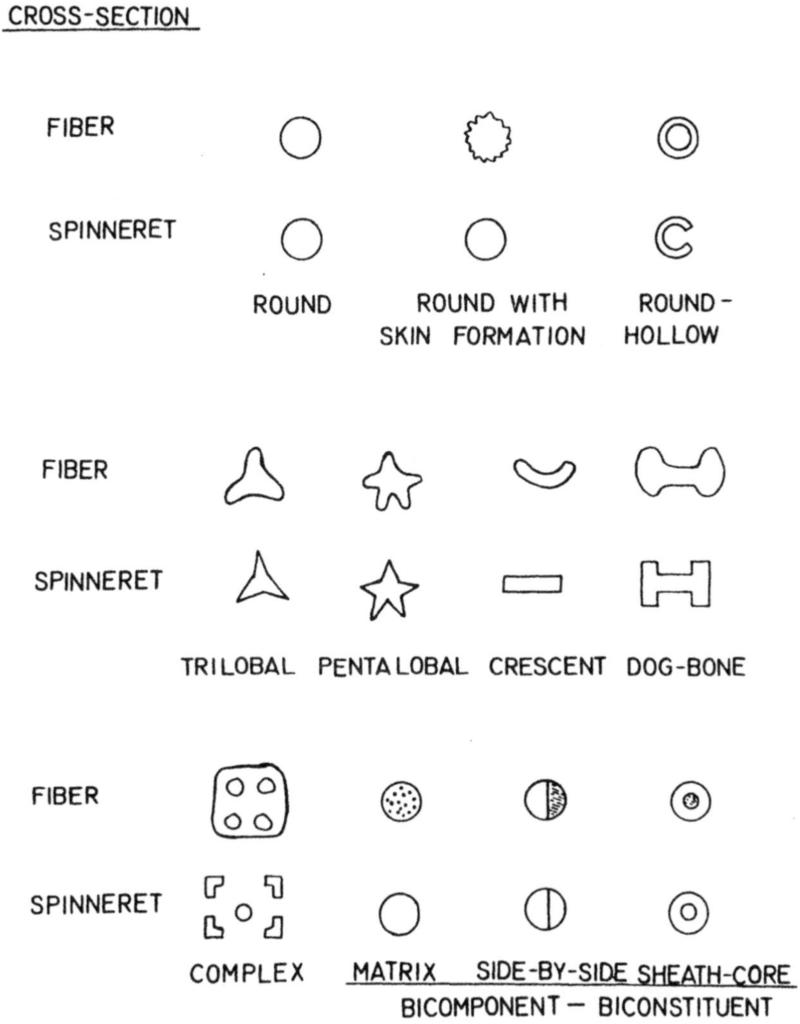

CROSS-SECTION

FIBER

SPINNERET

ROUND   ROUND WITH SKIN FORMATION   ROUND-HOLLOW

FIBER

SPINNERET

TRILOBAL   PENTALOBAL   CRESCENT   DOG-BONE

FIBER

SPINNERET

COMPLEX   MATRIX   SIDE-BY-SIDE   SHEATH-CORE

BICOMPONENT — BICONSTITUENT

Figure 1-5. Fiber cross sections.

A number of theories exist concerning the arrangement of crystalline and amorphous areas within a fiber. Individual crystalline areas in a fiber are often referred to as microfibrils. Microfibrils can associate into larger crystalline groups, which are called fibrils or micelles. Microfibrils are 30-100 Å (10^{-10} meters) in length, whereas fibrils and micelles are usually 200-600 Å in length. This compares to the individual molecular chains, which vary from 300 to 1500 Å in length and which are usually part of both crystalline and amorphous areas of the polymer and therefore give continuity and association of the various crystalline and amorphous areas within the fiber. A number of theories have been developed to explain the interconnection of crystalline and amorphous areas in the fiber and include such concepts as fringed micelles or fringed fibrils, molecular chain foldings, and extended chain concepts. The amorphous areas within a fiber will be relatively loosely packed and associated with each other, and spaces or voids will appear due to discontinuities within the structure. Figure 1-6 outlines the various aspects of internal fiber morphology with regard to polymer chains.

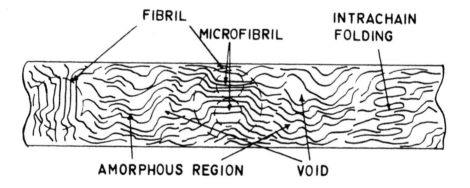

Figure 1-6. Aspects of internal fiber morphology.

The forces that keep crystalline areas together within a fiber include chemical bonds (covalent, ionic) as well as secondary bonds (hydrogen bonds, van der Waals forces, dipole-dipole interactions). Covalent bonds result from sharing of electrons between atoms, such as found in carbon-carbon, carbon-oxygen, and carbon-nitrogen bonding, within organic compounds. Covalent bonds joining adjacent polymer chains are referred to as crosslinks. Ionic bonding occurs when molecules donate or accept electrons from each other, as when a metal salt reacts with acid side chains on a polymer within a fiber. Chemical bonds are much stronger than secondary bonds formed between polymer chains, but the total associative force between polymer chains can be large, since a very large number of such bonds may occur between adjacent polymer chains. Hydrogen bonds are the strongest of the secondary bonds and occur between electropositive hydrogen atoms and electronegative atoms such as oxygen, nitrogen, and halogens on opposing polymer chains. Nylon, protein, and cellulosic fibers are capable of extensive hydrogen bonding. Van der Waals interactions between polymer chains occur when clouds of electrons from each chain come in close proximity, thereby promoting a small attractive force between chains. The more extended the cloud of electrons, the stronger the van der Waals interaction will be. Covalent bonded materials will show some uneven distribution of electron density over the molecule due to the differing electronegativity of the atoms and electron distribution over the molecule to form dipoles. Dipoles on adjacent polymer chains of opposite charge and close proximity are attracted to each other and promote secondary bonding.

When a synthetic fiber is stretched or drawn, the molecules in most cases will orient themselves in crystalline areas parallel to the fiber axis, although crystalline areas in some chain-folded polymers such as polypropylene can be aligned vertical to the fiber axis. The degree of crystallinity will be affected by the total forces available for chain interaction, the distance between parallel chains, and the similarity and uniformity of adjacent chains. The structure and arrangement of individual polymer chains also affects the morphology of the fiber. Also, cis-trans configurations or optical isomers of polymers can have very different physical and chemical properties.

Bulking, Texturizing, and Staple Formation

Thermoplastic man-made fibers can be permanently heat-set after drawing and orientation. The fiber will possess structural integrity and will not shrink up to that setting temperature. Also, thermoplastic fibers or yarns from these fibers can be texturized to give three dimensional loft and bulkiness (1) through fiber deformation and setting at or near their softening temperature, (2) through air entanglement, or (3) through differential setting within fibers or yarns (Table I). Schematic representations of these methods are given in Figure 1-7.

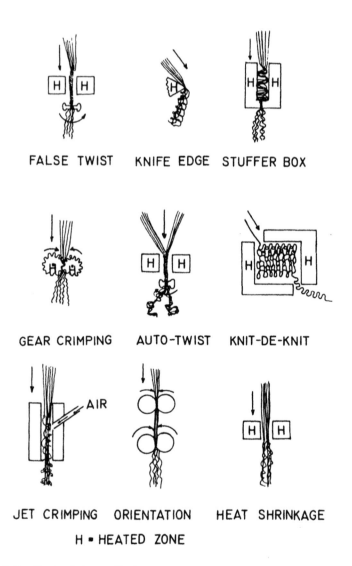

FALSE TWIST   KNIFE EDGE STUFFER BOX

GEAR CRIMPING   AUTO-TWIST   KNIT-DE-KNIT

JET CRIMPING   ORIENTATION   HEAT SHRINKAGE

H = HEATED ZONE

Figure 1-7. Texturizing methods.

TABLE I.   TEXTURIZING METHODS

Heat-setting Techniques	Air Entanglement	Differential Setting
False twist	Air jet	Bicomponent-biconstituent fiber orientation
Knife edge		
Stuffer box		Heat shrinkage of thermoplastic fibers in a blend
Gear crimping		
Autotwist		
Knit-de-knit		

Heat-Setting Techniques: The false twist heat-setting technique is extremely rapid, inexpensive, and most widely used. The filament fiber tow is brought in contact with a high-speed spindle running vertical to the moving tow. This action results in a high twist in the tow up to the spindle. The twisted tow is heated near its softening point before passing the spindle and then cooled and untwisted to give a wavy bulky yarn. In knife edge texturizing, filament tows or yarns are passed over a heated knife edge under tension. The fibers near the knife edge are changed in overall orientation in relation to the unheated yarns or portion of the filaments away from the knife edge, thereby causing bulking of the yarn. In stuffer box texturizing, the filament tow is fed into a heated box, causing the tow to double up against itself. On removal, the cooled tow retains the zigzag configuration caused by the process. In gear crimping, the tow is passed between heated intermeshing gears. On cooling the fibers retain the shape induced by the heated gears. In autotwisting, two tows or yarns are twisted together and then heat set. On untwisting the yarns have equal but opposite twists, which causes a spiral bulking of the yarn. In the knit-de-knit process a yarn is fill knitted, heat set, cooled, and de-knitted to give a bulked yarn retaining the shape and curvature of the knit.

Air Entanglement: In air entanglement texturizing, a fiber tow is loosely fed into and through a restricted space and a high-speed air jet is impinged on the fibers at a 45° angle. The loose fibers within the tow are looped to give a texturized effect.

Differential Setting: Heat shrinkage techniques cause a bulking of fiber tows containing different fibers through heating one component of the blend sufficiently to cause heat shrinkage of the fiber and compaction, contraction, and bulking. Side-by-side bicomponent and biconstituent fibers recover different degrees on each side from fiber stretching and causing a waving, crimping, or bulking of the fiber.

Staple Formation

Continuous filaments can be cut into staple by wet or dry cutting techniques. In wet cutting, the wet spun fiber is cut to uniform lengths right after spinning, while dry cutting involves partial cutting, debonding, and shuffling of the dry tow to form a sliver.

Before the filament or staple is used in yarn spinning, spin finishes are added to give lubricity and antistatic characteristics to the fibers and to provide a greater degree of fiber cohesiveness. Such finishes are usually mixtures including such materials as fatty acid esters, mineral oils, synthetic esters, silicones, cationic amines, phosphate esters, emulsifiers, and/or nonionic surfactants. Spin finishes are formulated to be oxidation resistant, to be easily removed by scouring, to give a controlled viscosity, to be stable to corrosion, to resist odor and color formation, and to be non-volatile and readily emulsifiable.

STRUCTURE-PROPERTY RELATIONSHIPS

The basic chemical and morphological structure of polymers in a fiber determine the fundamental properties of a fabric made from that fiber. Although physical and chemical treatments and changes in yarn and fabric formation parameters can alter the fabric properties to some degree, the basic properties of the fabric result from physical and chemical properties inherent to the structure of the polymer making up the fiber. From these basic properties, the end-use characteristics of the fiber are determined. To that end, in the following chapters we will describe the various textile fibers in terms of their basic structural properties, followed by physical and chemical properties, and finally the end-use characteristics inherent to constructions made from the fiber.

Initially the name and general information for a given fiber is set forth followed by an outline of the structural properties, including information about chemical structure of the polymer, degree of polymerization, and arrangement of molecular chains within the fiber. Physical properties include mechanical (tensile) and environmental properties of the fiber, whereas the effect of common chemicals and chemically induced processes on the fiber is listed under chemical properties. The end-use properties are then listed and include properties coming inherently from the structural, physical, and chemical properties of the fiber as well as end-use properties that involve evaluation of performance, subjective aspects, and aesthetics of the fabrics. Where possible, the interrelationships of these properties are presented.

2. Fiber Identification and Characterization

FIBER IDENTIFICATION

Several methods are used to identify fibers and to differentiate them from one another. The most common methods include microscopic examination, solubility tests, heating and burning characteristics, density or specific gravity, and staining techniques.

Microscopic Identification

Examination of longitudinal and cross-sectional views of a fiber at 100 to 500 magnifications gives detailed information with regard to the surface morphology of the fiber. Positive identification of many natural fibers is possible using the microscope, but positive identification of man-made fibers is more difficult due to their similarity in appearance and due to the fact that spinning techniques and spinneret shape can radically alter the gross morphological structure of the fiber.

Solubility

The chemical structure of polymers in a fiber determines the fiber's basic solubility characteristics, and the effect of solvents on fibers can aid in the general fiber classification. Various classification schemes involving solubility have been developed to separate and identify fibers.

Heating and Burning Characteristics

The reaction of fibers to heat from an open flame is a useful guide in identification of fibers. When thermoplastic fibers are brought close to a flame, they will melt, fuse, and shrink, whereas nonthermoplastic fibers will brown, char, or be unaffected by the flame. On contact with an open flame, fibers of organic polymers will ignite and burn. The nature of the burning reaction is characteristic of the chemical structure of the fiber. On removal from the flame, fibers will either self-extinguish or continue to burn. The odor of gases coming from the decomposing fibers and the nature of any residual ash are characteristic of the fibrous polymer being burned.

Density or Specific Gravity

Fiber density may be used as an aid in fiber identification. Fiber density may be determined by using a series of solvent mixtures of varying density or specific gravity. If the specific gravity of the fiber is greater than that of the liquid, the fiber specimen will sink in the liquid. Conversely, if the specific gravity of the fiber is less than that of the liquid, the fiber specimen will float. Thereby an approximate determination of fiber density may be made.

Staining

Fibers have differing dyeing characteristics and affinities dependent on the chemical and morphological structure of the fiber. Prepared dye mixtures containing dyes of differing affinities for various fiber types have been used extensively as identification stains for undyed fabrics. Since some fiber types may dye to similar shades with these dye mixtures, two or more stains usually must be used to confirm the fiber content. Staining is effective only for previously undyed fibers or for fibers where the dye is stripped from the fiber prior to staining.

STRUCTURAL, PHYSICAL, AND CHEMICAL CHARACTERIZATION

A number of methods are available for characterization of the structural, physical, and chemical properties of fibers. The major methods available are outlined in this chapter, including a brief description of each method and the nature of characterization that the method provides.

Optical and Electron Microscopy

Optical microscopy (OM) has been used for many years as a reliable method to determine the gross morphology of a fiber in longitudinal as well as cross-sectional views. Mounting the fiber on a slide wetted with a liquid of appropriate refractive properties has been used to minimize light scattering effects. The presence of gross morphological characteristics such as fiber shape and size and the nature of the surface can be readily detected. Magnifications as high as x1500 are possible, although less depth of field exists at higher magnifications. Scanning electron micros-copy (SEM) can be used to view the morphology of fibers with good depth of field and resolution at magnifications up to x10,000. In scanning electron microscopy, the fiber must first be coated with a thin film of a conducting metal such as silver or gold. The mounted specimen then is scanned with an electron beam, and particles emitted from the fiber surface are detected and analyzed to form an image of the fiber. Transmission electron micros-copy (TEM) is more specialized and more difficult to perform than SEM. It measures the net density of electrons passing through the thin cross sections of metal-coated fibers and provides a method to look at the micro-morphology of a fiber.

Elemental and End-Group Analysis

The qualitative and quantitative analysis of the chemical elements and groups in a fiber may aid in identification and characterization of a fiber. Care must be taken in analysis of such data, since the presence of dyes or finishes on the fibers may affect the nature and content of ele-ments and end-groups found in a given fiber. Gravimetric and instrumental chemical methods are available for analysis of specific elements or groups of elements in fibers. Specific chemical analyses of functional groups and end-groups in organic polymers that make up fibers may be carried out. For example, analyses of amino acids in protein fibers, amino groups in poly-amides and proteins, and acid groups in polyamides and polyesters aid in structure determination, molecular characterization, and identification of fibers.

Infrared Spectroscopy

Infrared spectroscopy is a valuable tool in determination of function-al groups within a fiber. Functional groups in a polymer absorb infrared energy at wavelengths characteristic of the particular group and lead to changes in the vibrational modes within the functional group. As a result of the infrared absorption characteristics of the fiber, specific functional groups can be identified. Infrared spectroscopy of fibers can be carried out on the finely divided fiber segments pressed in a salt pellet or through the use of reflectance techniques. Functional groups in dyes and finishes also can be detected by this technique.

Ultraviolet-Visible Spectroscopy

The ultraviolet-visible spectra of fibers, dyes, and finishes can provide clues concerning the structure of these materials, as well as show the nature of electronic transitions that occur within the material as light is absorbed at various wavelengths by unsaturated groups giving an electronically excited molecule. The absorbed energy is either harmlessly dissipated as heat, fluorescence, or phosphorescence or causes chemical reactions to occur that modify the chemical structure of the fiber. Ultraviolet-visible spectra can be measured for a material either in solution or by reflectance. Reflectance spectra are particularly useful in color measurement and assessment of color differences in dyed and bleached fibers.

Nuclear Magnetic Resonance Spectroscopy

Nuclear magnetic resonance (NMR) spectroscopy measures the relative magnitude and direction (moment) of spin orientation of the nucleus of the individual atoms within a polymer from a fiber in solution in a high-intensity magnetic field. The degree of shift of spins within the magnetic field and the signal splitting characteristics of individual atoms such as hydrogen or carbon within the molecule are dependent on the location and nature of the groups surrounding each atom. In this way, the "average" structure of long polymeric chains can be determined. Line width from NMR spectra also can provide information concerning the relationship of crystalline and amorphous areas within the polymer.

X-Ray Diffraction

X-rays diffracted from or reflected off of crystalline or semicrystalline polymeric materials will give patterns related to the crystalline and amorphous areas within a fiber. The size and shape of individual crystalline and amorphous sites within the fiber will be reflected in the geometry and sharpness of the x-ray diffraction pattern and will provide an insight into the internal structure of polymeric chains within a fiber.

Thermal Analysis

Physical and chemical changes in fibers may be investigated by measuring changes in selected properties as small samples of fiber are heated at a steady rate over a given temperature range in an inert atmosphere such as nitrogen. The four thermal characterization methods are (1) differential thermal analysis (DTA), (2) differential scanning calorimetry (DSC), (3) thermal gravimetric analysis (TGA), and (4) thermal mechanical analysis (TMA).

In DTA, small changes in temperature (ΔT) in the fiber sample compared to a reference are detected and recorded as the sample is heated. The changes in temperature (ΔT) are directly related to physical and chemical events occurring within the fiber as it is heated. These events include changes in crystallinity and crystal structure, loss of water, solvents, or volatile materials, melting, and decomposition of the fiber. DSC is similar to DTA but measures changes in heat content (ΔH) rather than temperature (ΔT) as the fiber is heated, and it provides quantitative data on the thermodynamic processes involved. In an inert gas such as nitrogen, most processes are endothermic (heat absorbing). If DTA or DSC is carried out in air with oxygen, data may be obtained related to the combustion characteristics of the fiber, and fiber decomposition becomes exothermic (heat generating). TGA measures changes in mass (ΔM) of a sample as the temperature is raised at a uniform rate. TGA provides information concerning loss of volatile materials, the rate and mode of decomposition of the fiber, and the effect of finishes on fiber decomposition. TMA measures changes in a specific mechanical property as the temperature of the fiber is raised at a uniform rate. A number of specialized mechanical devices have been developed to measure mechanical changes in fibers, including hardness and flow under stress.

Molecular Weight Determination

Molecular weight determination methods provide information concerning the average size and distribution of individual polymer molecules making up a fiber. Molecular weights enable one to calculate the length of the average repeating unit within the polymer chain, better known as the degree of polymerization (DP). The distribution of polymer chain lengths within the fiber provides information concerning selected polymer properties.

The major molecular weight determination methods include number average molecular weights ($\bar{M}_n$), determined by end-group analysis, osmometry, cryoscopy, and ebullioscopy; weight average molecular weights ($\bar{M}_w$), determined by light scattering and ultracentrifugation; and viscosity molecular weights ($\bar{M}_v$), determined by the flow rate of polymer solutions. Since each method measures the average molecular weight of the polymer differently, the molecular weight values obtained will differ depending on the overall number and distribution of polymer chains of varying lengths present in the fiber. The differences in value between $\bar{M}_n$ and $\bar{M}_w$ provide measures of the breadth of distribution of polymers within the fiber. By definition the distribution of molecular weights for a given polymer will always be $\bar{M}_w > \bar{M}_v > \bar{M}_n$.

Mechanical and Tensile Property Measurements

Mechanical and tensile measurements for fibers include tenacity or tensile strength, elongation at break, recovery from limited elongation, stiffness (relative force required to bend the fiber), and recovery from bending. The tensile properties of individual fibers or yarns are usually measured on a tensile testing machine such as an Instron, which subjects fibers or yarns of a given length to a constant rate of force or loading. The force necessary to break the fiber or yarn, or tenacity, is commonly given in grams per denier (g/d) or grams per tex (g/tex), or as kilometer breaking length in the SI system. The elongation to break of a fiber is a measure of the ultimate degree of extension that a fiber can withstand before breaking. The degree of recovery of a fiber from a given elongation is a measure of the resiliency of the fiber to small deformation forces. The stiffness or bendability of a fiber is related to the overall chemical structure of the macromolecules making up the fiber, the forces between adjacent polymer chains, and the degree of crystallinity of the fiber. Mechanical and tensile property measurements can provide valuable insights into the structure of a fiber and its projected performance in end-use.

Specific Gravity

The specific gravity of a fiber is a measure of its density in relation to the density of the same volume of water, and provides a method to relate the mass per unit volume of a given fiber to that of other fibers. Specific gravity relates to some degree to the nature of molecular packing, crystallinity, and molecular alignment in the fiber. Specific gravity of a fiber will give an idea of the relative weight of fabrics of identical fabric structure but of differing fiber content. End-use properties such as hand, drapability, and appearance are affected by fiber density.

Environmental Properties

Environmental properties include those physical properties which relate to the environment in which a fiber is found. Moisture regain, solvent solubility, heat conductivity, the physical effect of heat, and the electrical properties depend on the environmental conditions surrounding the fiber. The uptake of moisture by a dry fiber at equilibrium will depend on the temperature and relative humidity of the environment. Solvent solubilities of fibers will depend on the solubility parameters of the solvent in relation to fiber structure and crystallinity. Heat conductivity, the physical effect of heating such as melting, softening, and other thermal transitions, and electrical properties of a fiber depend on the inherent structure of the fiber and the manner in which heat or electrical energy is acted upon by the macromolecules within the fiber. Environmental properties are measured by subjecting the fiber to the appropriate environmental conditions and measuring the property desired under such conditions.

Chemical Properties

The chemical properties of fibers include the effects of chemical agents including acids, bases, oxidizing agents, reducing agents, and biological agents such as molds and mildews on the fiber and light- and heat-induced chemical changes within the fiber. Acids and bases cause hydrolytic attack of molecular chains within a fiber, whereas oxidizing and reducing agents will cause chemical attack of functional groups through oxidation (removal of electrons) or reduction (addition of electrons). Such chemical attack can change the fiber's structure and possibly cleave the molecular chains within the fiber. Biological agents such as moths on wool or mildew on cellulosics use the fiber as a nutrient for biological growth and subsequently cause damage to the fiber structure.

Sunlight contains ultraviolet, visible, and infrared light energy. This energy can be absorbed at discrete wavelength ranges by fibers depending on the molecular structure of the fiber. Ultraviolet and visible light absorbed by a fiber will cause excitation of electrons within the structure, raising them to higher energy states. Shorter ultraviolet wavelengths are the most highly energetic and give the most highly excited states. Visible light usually has little effect on the fiber, although its absorption and reflectance of unabsorbed light will determine the color and reflectance characteristics of the fiber. Infrared energy absorbed will increase the vibration of molecules within the fiber and causes heating of the fiber. The excited species within the fiber can return to their original (ground) state through dissipation of the energy as molecular vibrations or heat without significantly affecting the fiber. Ultraviolet and some visible light absorbed by the fiber, however, can lead to molecular scission within the fiber and cause adverse free radical reactions, which will lead to fiber deterioration.

Heating a fiber to progressively higher temperatures in air will lead to physical as well as chemical changes within the fiber. At sufficiently high temperatures, molecular scission, oxidation, and other complex chemical reactions associated with decomposition of the fiber will occur causing possible discoloration and a severe drop in physical and end-use properties for the fiber.

END-USE PROPERTY CHARACTERIZATION

End-use property characterization methods often involve use of laboratory techniques which are adapted to simulate actual conditions of average wear on the textile or that can predict performance in end-use. Often quantitative numerical values cannot be listed in comparing the end-use properties of a given textile fiber; nevertheless, relative rankings are possible and can give useful information about the suitability of a fabric from a given fiber type for a specific application. It must be emphasized that one must be most careful in interpreting results from test methods and extrapolating the findings to actual wear and use conditions. The ultimate properties of fibers in end-use do reflect the underlying morphological, physical, and chemical characteristics inherent to the fiber. All major end-use properties and characteristics considered in this handbook are outlined below. End-use methods are usually voluntary or mandatory standards developed by test or trade organizations or by government agencies. Organizations involved in standards development for textile end-use include the following:

American Association of Textile Chemists and Colorists (AATCC)
American National Standards Institute (ANSI)
American Society for Testing and Materials (ASTM)
Consumer Product Safety Commission (CPSC)
Federal Trade Commission (FTC)
Society of Dyers and Colorists (SDC)
International Standards Organization (ISO)

Characteristics Related to Identity, Aesthetics, and Comfort

Fibers are known by common, generic, and trade names. The Textile Fiber Products Identification Act, administered by the Federal Trade Commission, established generic names for all major classes of fibers based on the structure of the fiber. Common natural fibers often are also designated by their variety, type, or country of origin, whereas man-made fibers manufactured by various firms are designated by trade name. Nevertheless, the labeled textile must include the generic name of the fiber(s) and the percentage content of each fiber within the textile substrate. Often trade names are selected which will convey to the consumer a particular "feeling," property, or use for that fiber. Nylon is an example of a trade name (selected by duPont for their polyamide fiber) which came into such common usage that it eventually was designated as the generic name of this fiber class by the Federal Trade Commission (FTC). As new fibers of novel structure are developed and commercialized, new generic names are designated by the FTC.

A number of fiber end-use properties in textile constructions relate to the aesthetic, tactile, and comfort characteristics of the fiber. Such properties include appearance, luster, hand (feel or touch), drapability, absorbency, overall comfort, crease retention, pilling, and wrinkle resistance. All of these factors are affected to varying extents by the particular properties desired from the textile structure and its intended use. Many of these properties are related to inherent properties of the fibers, which are translated into textile structures prepared for end-use.

The overall appearance and luster of a textile can be related to the shape and light absorbing and scattering characteristics of the individual fiber within the structure. The "hand" or "handle" or a textile structure is a complex synthesis of tactical responses by an individual, and is characteristic of the particular fiber or fiber blend and overall structure of the textile substrate. The drapability of textiles is related to the fiber stiffness and bendability within the complex structural matrix making up the textile. The moisture absorbency and comfort of a fiber is related to its chemistry and morphology and to the way it absorbs, interacts with, and conducts moisture. In addition, comfort is related to the yarn and fabric structure into which the individual yarns have been made. Crease retention and wrinkle resistance of a fiber in a textile construction are directly related to the inherent chemical and morphological characteristics of the fiber as they depend on deformation and recovery under dry and moist conditions. The pilling characteristics of a fiber in a textile construction are related to the ease with which individual fibers may be partially pulled from the textile structure and to the tenacity of the individual fibers. Fibers in loose, open textile structure are readily pulled from the textile. If the fiber is strong, the fiber tangles with other loose fibers and mixes with lint and fiber fragments to form a pill. Weaker fibers such as cotton, however, usually break off before pill formation occurs.

Characteristics Related to Durability and Wear

The useful life of a fabric depends on a number of factors, including the strength, stretch, recovery, toughness, and abrasion resistance of the fiber and the tearing and bursting resistance of the fabrics made from that fiber. The composite of these factors coupled with the conditions and type of end-use or wear will determine the durability characteristics of a textile structure made from the fiber.

Fibers must be of minimum strength in order to construct textile structures with reasonable wear characteristics. The wear and durability of a fabric will tend to increase with increasing fiber strength. Textile structures made from fibers able to withstand stretching and deformation with good recovery from deformation will have improved durability particularly when subjected to bursting or tearing stresses. The relative toughness of the fiber also will affect the fabric durability, with tougher fibers giving the best performance. Tough but resilient fibers will also be resistant to abrasion or wear by rubbing the fiber surface. Abrasion of a textile structure usually occurs at edges (edge abrasion), on flat surfaces (flat abrasion), or through flexing of the textile structure resulting in interfiber abrasion (flex abrasion).

Physical and Chemical Characteristics and Response of Fiber to Its Environmental Surroundings

The physical and chemical characteristics of a fiber affect a number of important end-use properties: (1) heat (physical and chemical) effect on fibers, including the safe ironing temperature and flammability, (2) wetting of and soil removal from the fiber, including laundering, drycleaning, and fiber dyeability and fastness, and (3) chemical resistance, including resistance to attack by household chemicals and atmospheric gases, particularly in the presence of sunlight.

Fibers respond to heat in different ways. Thermoplastic fibers such as polyesters soften and eventually melt on heating without extensive decomposition, thereby permitting setting of the softened fiber through stretching and/or bending and subsequent cooling. Other fibers, such as the cellulosics and protein fibers, decompose before melting and therefore cannot be set using physical means. The safe ironing temperature of a fabric is determined by the softening and/or decomposition temperature of the fiber and must be significantly below this temperature. At sufficiently high temperatures, fibers are chemically attacked by oxygen in the atmosphere which accelerates fiber decomposition. If the temperature and heat input is sufficiently high or if a flame is involved, the fiber will ignite and burn and thereby decompose at a more rapid rate. On removal from the heat source, some fibers will self-extinguish, whereas others will sustain flaming combustion and continue to burn. The burning characteristics of a fiber depends on its inherent chemical structure and the nature of any finishes or additives present on the fiber.

When soil is removed from a fabric as in laundering or dry-cleaning, the individual fibers must be resistant to attack or damage caused by additives such as detergents, the solvent medium used, and mechanical agitation. Fabrics constructed of fibers that swell in water or dry-cleaning solvents can undergo profound dimensional changes on wetting. Also, fibers with surface scales such as wool undergo felting in the presence of moisture and mechanical action.

The dyeability of a fiber is dependent on the chemical and morphological characteristics of the fiber, the ability of the fiber to be effectively wetted and penetrated by the dyeing medium, and the diffusion characteristics of the dye in the fiber. Since most dyeing processes are done in water medium, hydrophilic fibers generally dye more readily than the more hydrophobic fibers. The fastness of the dye on the fiber will be dependent on the nature and order of physical and/or chemical forces holding the dye on the fiber and the effect of environmental factors such as sunlight, household chemicals, and mechanical action (crocking) on the dye-fiber combination.

The chemical resistance of a fiber can have a profound effect on end-use. The fibers that are sensitive to chemical attack by household chemicals such as bleach are limited in their end-uses. The resistance of fibers to attack by atmospheric gases including oxygen, ozone, and oxides of nitrogen, particularly in the presence of sunlight and moisture, can also be important considerations in certain end-uses.

II. Fiber Properties

3. Cellulosic Fibers

Cellulose is a fibrous material of plant origin and the basis of all natural and man-made cellulosic fibers. The natural cellulosic fibers include cotton, flax, hemp, jute, and ramie. The major man-made cellulosic fiber is rayon, a fiber produced by regeneration of dissolved forms of cellulose. The cellulose acetates are organic esters of cellulose and will be discussed in Chapter 4.

Cellulose is a polymeric sugar (polysaccharide) made up of repeating 1,4-β-anhydroglucose units connected to each other by β-ether linkages. The number of repeating units in cellulosic fibers can vary from less than 1000 to as many as 18,000, depending on the fiber source. Cellulose is a hemiacetal and hydrolyzes in dilute acid solutions to form glucose, a simple sugar.

The long linear chains of cellulose permit the hydroxyl functional groups on each anhydroglucose unit to interact with hydroxyl groups on adjacent chains through hydrogen bonding and van der Waals forces. These strong intermolecular forces between chains, coupled with the high linearity of the cellulose molecule, account for the crystalline nature of cellulosic fibers. It is believed that a gradual transition from alternating areas of greater molecular alignment or crystallinity to more disordered or amorphous areas occurs in cellulose. The number, size, and arrangement of crystalline regions within celluloses determine the ultimate properties of a particular fiber.

$$\left[\begin{array}{c} \text{CH}_2\text{OH} \\ \text{HO} \quad \text{HO} \quad \text{O} \end{array}\right]_n$$

CELLULOSE

The predominant reactive groups within cellulose are the primary and secondary hydroxyl functional groups. Each repeating anhydroglucose unit contains one primary and two secondary hydroxyl functional groups which are capable of undergoing characteristic chemical reactions of hydroxyl groups. The primary hydroxyls are more accessible and reactive than secondary hydroxyls; nevertheless, both types enter into many of the chemical reactions characteristic of cellulose.

COTTON

Cotton is the most important of the natural cellulosic fibers. It still accounts for about 50% of the total fiber production of the world, although man-made fibers have made significant inroads into cotton's share during the last three decades. Cotton fibers grow in the seed hair pod (boll) of cotton plants grown and cultivated in warm climates.

Structural Properties

Cotton is very nearly pure cellulose. As many as 10,000 repeating an-hydroglucose units are found in the polymeric cellulosic chains of cotton. Studies have shown that all of the hydroxyl hydrogens in cotton are hydrogen bonded. These hydrogen bonds will hold several adjacent cellulose chains in close alignment to form crystalline areas called microfibrils. These microfibrils in turn align themselves with each other to form larger crystalline units called fibrils, which are visible under the electron microscope. In cotton the fibrils are laid down in a spiral fashion within the fiber. Modern fiber theory suggests that each cellulose molecule is present within two or more crystalline regions of cellulose will be held together. Between the crystalline regions in cotton, amorphous unordered regions are found. Voids, spaces, and irregularities in structure will occur in these amorphous areas, whereas the cellulose chains in crystalline regions will be tightly packed. Penetration of dyestuffs and chemicals occurs more readily in these amorphous regions. Approximately 70% of the cotton fiber is crystalline. Individual cotton fibers are ribbonlike structures of somewhat irregular diameter with periodic twists or convolutions along the length of the fiber (Figure 3-1). These characteristic convolutions, as well as the cross-sectional shape of cotton are caused by collapse of the mature fiber on drying.

Three basic areas exist within the cross section of a cotton fiber. The primary outer wall or cuticle of cotton is a protective tough shell for the fiber, while the secondary wall beneath the outer shell makes up the bulk of the fiber. The fibrils within the secondary wall are packed alongside each other aligned as spirals running the length of the fiber. The lumen in the center of the fiber is a narrow canal-like structure running the length of the fiber. The lumen carries nutrients to the fiber during growth, but on maturity the fiber dries and the lumen collapses.

Physical Properties

The high crystallinity and associative forces between chains in cotton result in a moderately strong fiber having a tenacity of 2-5 g/d (18-45 g/tex). The hydrophilic (water-attracting) nature of cotton and the effect of absorbed water on the hydrogen bonding within cotton cause the tensile strength of cotton to change significantly with changes in moisture content. As a result, wet cotton is about 20% stronger than dry cotton. Cotton breaks at elongations of less than 10%, and the elastic recovery of cotton is only 75% after only 2% elongation.

Figure 3-1. Cotton. x1000.

Cotton is a relatively stiff fiber; however, wetting of the fiber with water plasticizes the cellulose structure, and the cotton becomes more pliable and soft. The resiliency of dry and wet cotton is poor, and many finishes have been developed to improve the wrinkle recovery characteristics of cotton.

Cotton is one of the more dense fibers and has a specific gravity of 1.54.

The hydroxyl groups of cotton possess great affinity for water, and the moisture regain of cotton is 7%-9% under standard conditions. At 100% relative humidity, cotton has 25%-30% moisture absorbency.

The heat conductivity of cotton is high, and cotton fabrics feel cool to the touch. Cotton has excellent heat characteristics, and its physical properties are unchanged by heating at 120°C for moderate periods. The electrical resistivity of cotton is low at moderate relative humidities, and the fiber has low static electricity buildup characteristics.

Cotton is not dissolved by common organic solvents. Cotton is swollen slightly by water because of its hydrophilic nature, but it is soluble only in solvents capable of breaking down the associative forces within the crystalline areas of cotton. Aqueous cuprammonium hydroxide and cupri-ethylenediamine are such solvents.

Chemical Properties

Cotton is hydrolyzed by hot dilute or cold concentrated acids to form hydrocellulose but is not affected by dilute acids near room temperature. Cotton has excellent resistance to alkalies. Concentrated alkali solutions swell cotton, but the fiber is not damaged.

The swelling of cotton by concentrated sodium hydroxide solution is used to chemically finish cotton by a technique called mercerization. In aqueous alkali, the cotton swells to form a more circular cross section and at the same time loses convolutions. If the cotton is held fast during swelling to prevent shrinkage, the cellulose fibers deform to give a fiber of smoother surface. After washing to remove alkali, followed by drying, the cotton fiber retains a more cylindrical shape and circular cross section. Although little chemical difference exists between mercerized and unmercerized cotton, mercerization does give a more reactive fiber with a higher regain and better dyeability (Figure 3-2).

Dilute solutions of oxidizing and reducing agents have little effect on cotton; however, appreciable attack by concentrated solutions of hydrogen peroxide, sodium chlorite, and sodium hypochlorite is found.

Most insects do not attack cotton; however, silverfish will attack cotton in the presence of starch. A major problem with cotton results from fungi and bacteria being able to grow on cotton. Mildews feed on hot moist cotton fibers, causing rotting and weakening of the fibers. Characteristic odor and pigment staining of the cotton occurs when mildews attack. Additives capable of protecting cotton are available and commercially applied to cotton fabrics used outdoors. These materials are often metal salts of

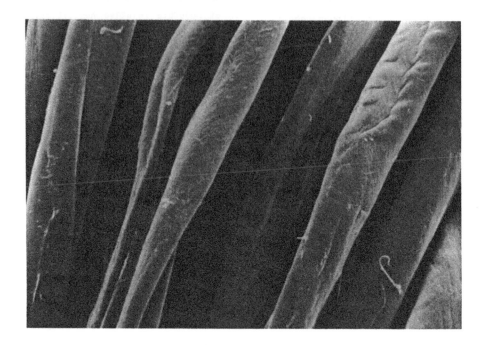

Figure 3-2. Mercerized Cotton. x1000.

organic compounds which are capable of inhibiting growth of mildews and similar organisms.

Cotton is only slowly attacked by sunlight, since cellulose lacks for the most part groups which absorb ultraviolet light between 300 and 400 nm. Over long periods sunlight degrades cotton, causing it to lose strength and to turn yellow. Certain vat dyes tend to accelerate the rate of cotton photodegradation through a sensitization reaction called "phototendering."

Although cotton has excellent heat resistance, degradation due to oxidation becomes noticeable when cotton is heated in the air at 150°C for long periods. Spontaneous ignition and burning of cotton occurs at 390°C. At low humidities in the absence of heat and light, cotton will not deteriorate over long periods of storage.

End-Use Properties

The properties of cotton fiber are such that it serves as nature's utility fiber. Although cotton has some properties which are undesirable from the viewpoint of the consumer, the superior properties of cotton, coupled with its low cost, nevertheless make it a valuable fiber in many applications. Different species of cotton produce fibers of various average lengths. In the United States lengths of cotton staples are designated as follows:

Extra long staple	1 3/8 - 2 inches (3.2 - 4.7 cm)
Long staple	1 1/8 - 1 3/8 inches (2.7 - 3.2 cm)
Medium staple	1 - 1 1/8 inches (2.5 - 2.7 cm)
Short staple	7/8 - 1 inch (2.0 - 2.5 cm)

The two major types of American cotton are American-Upland (including deltapine and acala varieties) and American-Egyptian (including pima).

Cotton has excellent hand, and the drapability of cotton fabrics is quite acceptable. Fabrics of cotton are of satisfactory appearance and have a low luster unless mercerized or resin finished.

The superior absorbency of cotton, coupled with its ability to desorb moisture, makes it a very comfortable fiber to wear. This absorbency permits cotton to be used in applications where moisture absorption is important, such as in sheets and towels. Mildewing of cotton under hot moist conditions and its slowness of drying are undesirable properties associated with its high affinity for water.

The cotton fiber has sufficient strength in the dry and wet states to make it suitable for most consumer textile applications. The increased strength of cotton on wetting adds to its long useful life. Cotton wears well without undue abrasion, and pills do not tend to form as it wears. Cotton's low resiliency and poor recovery from deformation means that it wrinkles easily in both the dry and wet states and exhibits inferior crease retention. Starching of cotton improves these properties, but the effect is only temporary, and it is necessary to renew this finish after each laundering.

The resistance of cotton to common household chemicals, sunlight, and heat makes it durable in most textile applications. Cotton can be dyed successfully by a wide variety of dyes, and the colorfastness of properly dyed cotton is satisfactory.

Fabrics of cotton are maintained with a moderate degree of care. Cotton fabrics launder readily, and its wet strength and alkali resistance mean that cotton is resistant to repeated washings. Stresses which occur during the spinning and weaving process will cause cotton fabrics to undergo relaxation shrinkage during initial launderings. Relaxation shrinkage can be controlled through resin stabilization or through the well known compression shrinkage process called Sanforization.

Cotton can be drip dried or tumbled dry, but in both cases the dry cotton will be severely wrinkled, and ironing will be necessary. Cotton can be ironed safely at temperatures as high as 205°C. Cotton, as well as all cellulosic fibers, is highly flammable and continues to burn after removal from a flame. After extinguishing of the flame the cotton will continue to glow and oxidize by a smoldering process called afterglow. The Limiting Oxygen Index (LOI) of cotton is 18. A number of topical treatments have been developed to lower the flammability of cotton and other cellulosics.

The undesirable properties of cotton can be corrected to varying degrees through treatment of the fiber with special finishes; however, the abrasion resistance of cotton is adversely affected. As a result, blends of cotton with the stronger man-made fibers have become important. Although man-made fibers have made inroads into applications previously reserved for cotton, cotton continues to be the major textile fiber due to its great versatility, availability, and cost.

FLAX

Flax is a bast fiber used to manufacture linen textiles. It is derived from the stem of the annual plant, Linum usitatissimum, which grows in many temperate and subtropical areas of the world. The flax plant grows as high as 4 feet tall, and flax fibers are found below the plant surface held in place by woody material and cellular matter. The flax fibers are freed from the plant by a fermentation process called retting. Chemical retting using acids and bases has been employed with some success; however, such processes tend to be more expensive than natural fermentation techniques. The fibers are then removed from the plant through breaking of the woody core, removal of the woody material (scutching), and combing (hackling). The resulting fibers are ready for spinning and are 12-15 inches (30-38 cm) in length.

Structural Properties

Flax is nearly pure cellulose and therefore exhibits many properties similar to cotton. Strands of individual flax fibers may consist of many individual fiber cells--fibrils held together by a natural cellulosic adhesive material. The molecular chains in flax are extremely long, and the average molecular weight of molecules in flax is 3 million. There are swellings or nodes periodically along the fiber which show up as characteristic cross markings. The cell walls of flax are thick, and the fiber cross section is polygonal with a large lumen in the center (Figure 3-3).

Physical Properties

Flax is a strong fiber, and the average tenacity of dry flax is 2-7 g/d (18-63 g/tex); wet flax is 2.5-9 g/d (23-81 g/tex). Like cotton, the wet flax fiber is about 20% stronger than the dry fiber. Flax has a low elongation at break (about 3%); however, the fiber is fairly elastic at very low elongations.

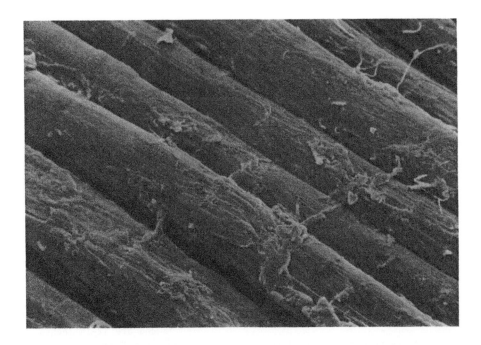

Figure 3-3. Flax. x1000.

Flax has a higher regain than cotton, and a 12% regain is observed at standard conditions. Flax possesses good heat conductivity and is cool to the touch. The heat and electrical properties of flax are similar to those of cotton, and flax possesses solubility characteristics identical to cotton.

Flax is a highly rigid fiber, and it tends to crease on bending due to its poor resiliency. The specific gravity of flax is 1.54, the same as cotton. Flax is a dull fiber, but more lustrous linen fabrics can be obtained by pounding the linen with wooden hammers (beetling). Linen fabrics have good dimensional stability unless stresses have been introduced during the weaving process.

Chemical Properties

The chemical properties of flax are similar to those of cotton. Flax has good resistance to acids, bases, and chemical bleaches. Flax is resistant to insects and microorganisms, and only under severe moist warm conditions will it be attacked by mildews. Flax is only slowly degraded by sunlight, and it decomposes at temperatures similar to that observed for cotton.

End-Use Properties

Flax finds only limited use in modern textiles. The methods used to produce flax involve extensive labor and cost; thus flax usage is reserved for special luxury applications. The strength of flax makes it superior to cotton in certain applications. Flax is resistant to abrasion, but like cotton suffers from lack of crease retention and poor wrinkle recovery. Linen fabric's superior absorbency, coupled with its cool crisp hand, contributes to its desirability as a prestige fiber. The thermal properties and chemical resistance of flax make it suitable for many consumer applications. Linen can be laundered repeatedly without deterioration and ironed safely at temperatures as high as 220°C. Flax is very flammable. Flax is used in prestige items including handkerchiefs, towels, tablecloths, sheets, and certain garments.

OTHER NATURAL CELLULOSIC FIBERS

A large number of natural cellulosic fibers from plant sources exist, including jute, hemp, ramie, kenaf, urena, sisal, henequen, abaca, pina, kapok, coir, and others. Each of the fibers is a plant or seed hair fiber and has properties that make it suitable for selected applications. The major fibers--jute, hemp, and ramie--will be considered here.

Hemp

Hemp is a bast fiber harvested from the hemp plant and processed in a manner similar to flax. It is a coarser fiber than flax, darker in color and difficult to bleach. The fiber is strong and durable, and the strands of hemp fiber may reach six feet or longer. Individual hemp cells are 1/2-1 inch (1.2-2.5 cm) in length, and the fiber cross section is polygonal. The fiber is very stiff and contains considerable lignin. Although fine fabrics can be produced from select hemps, hemp is used mainly in coarse fabrics, including sack material, canvas, ropes, and twines.

Jute

Jute fiber comes from a herbaceous annual plant which grows as high as 20 feet. The fibers are extracted from the plant stalk in a manner similar to flax. Jute tends to be brown in color due to about 20% lignin present in the fiber, but it does have a silklike luster. The individual cells of jute are very short, being about 0.1 inch (.25 cm) long, and the fiber cross section has five or six sides. Jute is not as strong or as durable as flax or hemp. Jute fibers are stiff, but it does have an unusually high moisture regain. Jute's low cost, moderate strength, and availability make it an important fiber for use in sacks, bags, and packing materials. Fabrics made of jute are called burlap.

Ramie

Ramie is a bast fiber often referred to as china grass. Ramie fiber is removed from the plant by peeling or removing the bark and soaking the fibers in water, followed by scraping. The fiber must be further degummed by treatment in base before spinning. Ramie is a white fiber with excellent strength and luster. The fiber is stiff and fairly coarse. The cells of the fiber are very long, and the cross section irregular in shape. Ramie is useful in industrial applications and is being used in furnishings where rough, irregular fabrics are desired.

RAYON

Rayon was the first man-made fiber to be produced commercially, and several types of rayon fibers are produced. Rayon is regenerated cellulosic material produced by solution of a cellulose source (wood pulp, cotton waste, etc.), followed by forcing of the solution through a spinneret and subsequent regeneration to form the fiber. The Federal Trade Commission defines rayon as a manufactured fiber composed of regenerated cellulose in which substituents have not replaced more than 15% of the hydroxyl hydrogens. Rayon is produced by three methods to give the viscose rayons, cuprammonium rayon, and saponified cellulose acetate. Of these, viscose rayon is the most important and the most inexpensive to produce.

Structural Properties and Formation of Rayon

Viscose Rayon: Viscose rayon fibers are produced by the viscose process, which involves formation of soda cellulose, reaction with carbon disulfide to form cellulose xanthate, and spinning into dilute acid to regenerate the cellulose as rayon fiber. In practice the cellulose pulp is steeped in warm concentrated sodium hydroxide and pressed to remove excess sodium hydroxide. After aging, during which some degradation and breaking of cellulose chains occurs, the soda cellulose crumbs are mixed with carbon disulfide, and orange sodium cellulose xanthate is formed. The xanthate is then dissolved in dilute sodium hydroxide. At this point titanium dioxide is added to the solution if a delustered fiber is desired. The solution is aged and ripened until a proper viscosity is reached; the solution is then forced through the spinneret into dilute sulfuric acid to decompose the xanthate and regenerate the cellulose by the wet spinning process. In addition to acid, the following additives are often found in the spinning bath: sodium sulfate (15%-20%), zinc sulfate (1%-5%), and glucose (2%). Initially, the outer portion of the xanthate decomposes in the acid bath and a cellulose skin forms on the fiber. The sodium and zinc sulfates affect the rate of xanthate decomposition and fiber formation, while the action of zinc sulfate is confined to crosslinking the outermost portion of the fiber. The glucose tends to slow the rate of regeneration of the fiber and tends to make the fiber more pliable. The concentration and type of additives in the bath and the nature of the original cellulose will affect the physical properties of the rayon. By careful control of the coagulation bath followed by mechanical stretching, crimp can be produced in the fiber. Slow regeneration and stretching of the rayon will tend to introduce greater areas of crystallinity within the fiber. High-tenacity rayons are formed in this manner. Also, the fiber can be modified through addition of other crosslinking agents to the spinning bath.

High-wet-modulus rayons (polynosic rayons) have been developed in recent years. These fibers are produced from high-grade cellulose starting materials, and the formation and decomposition of cellulose xanthate is carried out under the mildest of conditions to prevent degradation of the cellulose chains.

After initial skin formation, the core of the fiber decomposes, hardens, and shrinks, causing wrinkling on the fiber skin. The final fiber cross section of viscose rayon appears serrated and irregular, and the nature of the crystallites in the skin and core will differ (Figure 3-4). This skin effect can be controlled sufficiently to give nearly round fiber cross sections where carefully controlled coagulation occurs.

Figure 3-4. Viscose rayon. x2000.

The majority of viscose rayons have serrated and irregular cross sections due to the skin effect. The viscose rayon fiber is long and straight unless the fiber has been crimped, and striations due to the irregular cross section will run the length of the fiber. If the fiber has been delustered or spun dyed during fiber formation, particles of pigment will appear in the fiber. Normal viscose fibers will generally consist of 25%-30% crystalline areas within the fiber, and the average degree of polymerization of glucose units in the cellulose chains will be 200-700. The crystallites in viscose rayon are somewhat smaller than those found in cotton. Higher-tenacity rayons and high-wet-modulus rayons tend to be more crystalline, and higher degrees of polymerization are found. The cross sections of these rayons tend to be more nearly round. The degree of crystallinity of high-wet-modulus rayons can reach 55%.

Cuprammonium Rayon: Cuprammonium rayon is produced by solution of cellulosic material in cuprammonium hydroxide solution at low temperature under nitrogen, followed by extrusion of the solution through a spinneret into water and then sulfuric acid to decompose the cuprammonium complex and regenerate the cellulose. Cuprammonium rayon is more silklike than any of the other celluloses, but the cost of production is correspondingly higher. Cuprammonium rayon has a smooth surface, and no markings or striations are found on the fiber (Figure 3-5). The fiber cross section is nearly round.

Saponified Cellulose Acetate: Rayon can be produced from cellulose acetate yarns by hydrolysis in base, a process called saponification. Cellulose acetate molecules are unable to pack closely within the fiber due to lack of hydrogen bonding and the presence of bulky acetate side chains within the molecule. Since cellulose acetate is more plastic in nature, it can therefore be drawn and extended to give a high degree of orientation to the fiber. On hydrolysis the highly oriented cellulosic chains form hydrogen bonds with adjacent chains to give a rayon of high orientation and crystallinity. Saponified cellulose acetate has a nearly round but lobed cross section. The indentations caused by the lobes appear as light striations running the length of the filament.

Physical Properties

Since rayons are essentially pure cellulose, they would be expected to have physical properties similar to cellulosic fibers of natural origin. Differences in properties would be expected to depend on the degree of polymerization, crystallinity, and orientation within the fiber. The dry and wet tenacities of the rayons vary over a wide range dependent on the degree of polymerization and crystallinity with tenacities of 2 to 6 g/d (18 to 54 g/tex) dry and 1 to 4 g/d (9 to 36 g/tex) wet. For the more crystalline and oriented rayons, the drop in tenacity observed on wetting of the fiber is lower; however, none of the rayons exhibit higher wet strength than dry strength, as found with cotton. The great loss in strength of wet regular tenacity rayon makes it subject to damage during laundering.

Figure 3-5. Cuprammonium rayon. x1100.

The percentage elongation at break varies from 10 to 30% dry and 15 to 40% wet. The recovery at 2% elongation ranges from 70 to 100%. In general, the rayons have significantly higher elongations at break than observed for cotton. As the degree of crystallinity and orientation of the rayon increases, the elongation at break is lower. Rayons generally have somewhat better elastic properties at low elongations than is found for cotton.

The rayons possess greater luster than cotton, and often delusterants are added to rayon prior to spinning to give a more subdued fiber. Since the rayons are less crystalline and oriented than cotton, they tend to swell more in water and undergo greater elongation under tension in both the wet and dry states. During weaving and wet finishing considerable stresses can be introduced into many rayon fabrics, and relaxation of these stresses will be necessary before a dimensionally stable fabric is obtained.

Owing in part to the more extensive networks of amorphous regions found in rayons, the moisture regain of rayon is significantly higher than that of cotton. The regains of rayons under standard conditions range from 11% to 13%. The lower crystallinity and degree of polymerization of rayons also affect the way water acts on the fibers. Rayons as a consequence are swollen to varying degrees due to increased susceptibility to water penetration. During wetting rayon may increase up to 5% in length and swell up to double its volume.

The heat conductivity and electrical properties are the same as those found for cotton. Rayon is cool to the touch, and static charges do not build up in the fiber at humidities greater than 30%.

Although rayon swells in water, it is not attacked by common organic solvents. It does dissolve in cuprammonium solutions.

The rayons are moderately stiff fibers with poor resiliency and wrinkle recovery properties. As in the case of cotton, resin treatments will effectively increase the resiliency of rayon. Often such treatments will tend to be more effective on rayon than on cotton due to the greater accessibility of the interior of rayon to the resin. The specific gravities of viscose and cuprammonium rayons are the same as cotton and vary between 1.50 and 1.54. Only saponified cellulose acetate has a markedly different specific gravity (1.30).

Chemical Properties

The chemical properties of the rayons are essentially the same as those found for cotton. Cold dilute or concentrated base does not significantly attack rayon, but hot dilute acids will attack rayon at a rate faster than found for cotton. Although generally resistant to oxidizing bleaches, rayon is significantly attacked by hydrogen peroxide in high concentrations. Although silverfish are known to attack rayon, microorganisms which cause mildew do not readily attack rayon, except under more severe hot and moist conditions. Prolonged exposure of rayon to sunlight causes degradation of the cellulose chains and loss of strength of the fibers. The rayons are not thermoplastic, and they, like cotton, decompose below their melting point. Regular-tenacity rayon begins to decompose and lose strength at 150°C for prolonged periods and decomposes rapidly at 190°-210°C. High-tenacity rayons tend to decompose at slightly higher temperatures.

End-Use Properties

Rayon fibers found in consumer goods are known by numerous trade names. Regular- and high-tenacity viscose rayons are marketed as rayon or with names like Zantrel, Avril, Enkaire, and Fibro. Cuprammonium rayon and saponified cellulose acetate are not longer in production in the U.S. Regular- and medium-tenacity viscose rayons are among the least expensive of the man-made fibers, whereas high-strength, cuprammonium, and saponified cellulose acetate rayons are more expensive due to the greater care and additional steps necessary in manufacture of these fibers.

Although the properties of rayons are very nearly those of cotton, a greater range of properties exist within the various types of rayons available. Being an inexpensive fiber, rayon plays a role similar to that of cotton; however, rayon differs from cotton in that it can be modified to some degree during manufacture and is not as subject to world economic and climatic conditions.

The dry strength of regular-tenacity viscose and cuprammonium rayons are lower than that found for cotton, whereas high-tenacity viscose, polynosic, and saponified cellulose acetate rayons are significantly stronger than cotton. All rayons lose strength when wet and are more susceptible to damage while wet.

The abrasion resistance of rayon is fair, and abrasion of rayon fabrics becomes noticeable after repeated usage and launderings. The rayons are resistant to pill formation. Rayon fabrics wrinkle easily and without chemical treatment show poor crease recovery and crease retention. Durable press and wash-and-wear finishes tend to cause less degradation of rayon than found with cotton. Rayon possesses excellent moisture absorbency characteristics, but the weaker rayons swell and deform somewhat in the presence of moisture.

The hand of lower-strength viscose rayons is only fair, whereas the hand of cuprammonium and high-wet-modulus rayons is crisp and more desirable. Fabrics of rayon are generally comfortable to wear and are of acceptable appearance.

The rayons are resistant to common household solvents and to light and heat except under extreme conditions or prolonged exposures. Rayons can be dyed readily by a wide variety of dyes, and the colorfastness of dyed rayons is satisfactory. Rayons are moderately resistant to deterioration by repeated laundering, and a moderate degree of care is necessary for maintenance of rayon fabrics. The lower-strength rayons exhibit poor dimensional stability during laundering. Rayons are affected significantly by detergents or other laundry additives. Greater care must be taken when ironing rayon fabrics than would be necessary with cotton. Viscose and cuprammonium rayons may be safely ironed at 176°C, while saponified cellulose acetate can be ironed at 190°C. The rayons possess poor flammability characteristics and ignite readily on contact with a flame, as is the case for natural cellulosics.

The versatility of rayons, coupled with their lower price, makes them suitable for many textile applications. Rayon is used in clothing and home furnishings. Disposable nonwoven garments and products of rayon have been introduced to the consumer in recent years. The stronger rayons have been used in tire cord for several decades but have lost a significant portion of this important market in recent years. Rayon has been used more and more in blends with synthetic fibers, since rayon undergoes less degradation than cotton with durable press and wash-and-wear finishes.

4. Cellulose Ester Fibers

ACETATE AND TRIACETATE

The cellulose esters triacetate and acetate (secondary acetate) are the two major fibers of this type. The production of acetate fibers resulted from an attempt to find a new outlet for cellulose acetate used as aircraft "dope" in World War I. By 1921, the first acetate fibers were being produced. Although small quantities of cellulose triacetate fibers were produced before World War I, it was not until after 1954 that cellulose triacetate fibers were produced commercially in large quantities.

The Federal Trade Commission defines the acetate fibers as manufactured fibers in which the fiber-forming substance is cellulose acetate. Triacetate fibers are those cellulose fibers in which more than 92% of the hydroxyl groups are acetylated. The term secondary acetate has been used in the past in reference to the acetates being less than 92% acetylated. In recent years, these fibers have simply gone by the designation acetate. Although triacetate and acetate are very similar structurally, this slight difference in degree of esterification provides significant differences in properties for these fibers.

Structural Properties and Formation

The cellulose esters triacetate and acetate are formed through acetylation of cotton linters or wood pulp using acetic anhydride and an acid catalyst in acetic acid. At this point the cellulose is fully acetylated.

ACETATE

TRIACETATE

The triacetate is precipitated from solution by addition of water and dried. The cellulose triacetate is normally dissolved in methylene chloride to form about a 20% solution and forced through the spinneret into a stream of hot air to evaporate the solvent to form triacetate fibers by dry spinning. Acetate is formed by allowing the water-precipitated cellulose triacetate solution above to stand as long as 20 hours in aqueous solution. During this ripening process, some of the acetyl groups are hydrolyzed or removed from the cellulose. The acetate is then washed, dried, and dissolved in acetone to form the spinning "dope." The "dope" is forced through a spinneret into hot air to form the acetate fiber by dry spinning. Triacetate and acetate fibers can also be formed by wet spinning processes.

In triacetate, essentially all hydroxyl groups of the cellulose have been acetylated, whereas 2.3-2.5 hydroxyl groups per anhydroglucose unit have been acetylated in cellulose acetate. In both triacetate and acetate, hydrogen bonding between cellulose chains for the most part is eliminated. Also, the bulky acetate group discourages close tight packing of adjacent

cellulosic chains. As would be expected, these changes in molecular structure greatly affect the physical properties of triacetate and acetate.

Although limited hydrogen bonding occurs between molecules within acetates, triacetate is incapable of forming hydrogen bonds. Van der Waals
forces caused by the interaction of adjacent acetate and triacetate chains
are the major associative forces between the acetylated cellulose molecular
chains. The average number of acetylated anhydroglucose units in chains of
acetates and triacetates usually varies between 250 and 300 units.

The microscopic structures of triacetate and acetate are similar (Figures 4-1 and 4-2). The fiber cross section of each fiber is irregular,
with as many as five or six lobes. The longitudinal views of these fibers
show striations running the length of the fiber. Small particles of delusterant will be visible in these fibers if they have been added to the fiber
prior to spinning. Acetate and triacetate fibers are very similar in
appearance to the regular-tenacity viscose rayons.

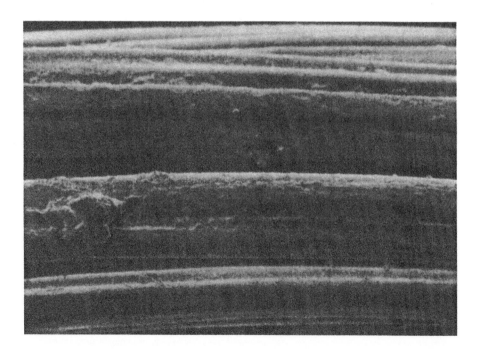

Figure 4-1. Acetate. x1200.

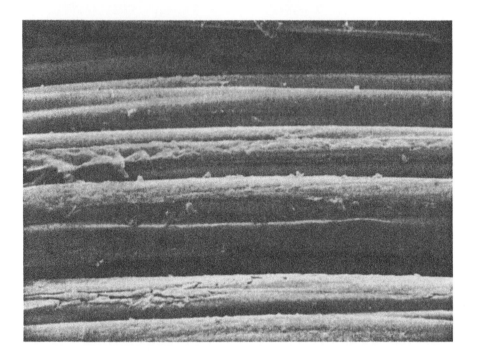

Figure 4-2. Triacetate. x1200.

Physical Properties

Since acetate and triacetate have only limited associative forces between molecular chains, they are considerably weaker than most of the cellulosic fibers. The tenacity of acetate and triacetate falls in the range 1-1.5 g/d (9-14 g/tex). On wetting, their tenacities fall to 0.8-1.2 g/d (7.2-10.8 g/tex). The elongations at break for acetate and triacetate are 25%-40%. On wetting, acetate breaks at 35%-50% elongation, whereas triacetate breaks at 30%-40% elongation. These fibers possess good recoveries at low elongations (80%-100% at 2% elongation). Above 5% elongation, plastic deformation of the fibers occurs and subsequent recovery is poor.

Acetates and triacetates are moderately stiff fibers and possess good resiliency on bending and deformation, particularly after heat treatment. The specific gravities of acetate and triacetate range from 1.30 to 1.35.

The moisture regains of these fibers are markedly lower than found for rayon and cotton. Acetate and triacetate have moisture regains of 6.0% and 4.5%, respectively. On heating, the moisture regain of triacetate drops further to 2.5%.

Acetate and triacetate fibers conduct heat readily and are cool to the touch. Being thermoplastic, acetate and triacetate fibers soften at 200°C and 225°C and melt at 232°C and 300°C, respectively. Triacetate can be heat set more readily than acetate, and permanent pleats can be satisfactorily heat set into triacetate fabric. Heat treatment of triacetate will change the character of the fiber and give a more dense molecular structure. Acetate and triacetate possess excellent electrical characteristics and can be used as insulating materials. As a consequence, static buildup is a problem on these fibers unless antistatic treatments are applied. Acetates and triacetates are attacked by a number of organic solvents capable of dissolving esters. Acetate swells or dissolves in acetone, ethyl acetate, phenol, chloroform, and methylene chloride, whereas triacetate dissolves in formic acid, acetic acid, methylene chloride, or chloroform.

Chemical Properties

Acetate and triacetate are resistant to dilute acids but are attacked by concentrated acids causing hydrolysis and removal of the acetate ester groups. While resistant to dilute alkalies, acetate is more susceptible to attack by alkalies than triacetate. Dilute alkali solutions attack acetate causing saponification and removal of the acetate groups on the fibers. Acetate is attacked by strong oxidizing agents, while triacetate has somewhat better resistance to attack by these reagents. Both fibers, however, can be bleached under proper conditions. Both acetate and triacetate are not attacked by insects or microorganisms. Acetate and triacetate are resistant to attack by sunlight, with greater resistance found for triacetate. Acetate and triacetate will withstand long periods at elevated temperatures below their melting points without significant loss of strength.

End-Use Properties

Acetates and triacetates are relatively inexpensive and have certain properties which are desirable in selected end-use applications. Acetate is marketed under trade names such as Airloft, Estrom, and Loftura, while triacetate fibers bear the trade names Tricel and Arnel.

Both acetate and triacetate are weak fibers having low tenacities, and they cannot be used in applications where high strength is required unless they are blended with other fibers. The high elongations at break found for acetate and triacetate help compensate for these fibers' low strengths to some extent, and the fibers exhibit good recoveries at low extensions.

The abrasion resistance of acetate and triacetate is poor, and these fibers cannot be used in applications requiring high resistance to rubbing and abrasion; however, the resistance of these fibers to pilling is excellent. While acetate and triacetate are moderately absorbent, their absorbencies cannot compare with the pure cellulosic fibers. The hand of acetate fabrics is somewhat softer and more pliable than triacetate, which possesses a crisp firm hand. Fabrics of both fibers possess excellent draping characteristics. Fabrics of acetate and triacetate have a pleasing appearance and a high degree of luster, but the luster of these fabrics can be modified through addition of delusterants.

Both acetate and triacetate are susceptible to attack by a number of household chemicals. Acetate and triacetate are attacked by strong acids and bases and by oxidizing bleaches. Acetate has only fair sunlight resistance, whereas the sunlight resistance of triacetate is superior. Both fibers have good heat resistance below their melting points.

Acetate and triacetate cannot be dyed by dyes used for cellulosic fibers. These fibers can be satisfactorily dyed with disperse dyes at moderate to high temperatures to give even, bright shades. The laundry colorfastness of triacetate is excellent; however, dyed acetate fabrics generally have moderate to poor colorfastness.

The acetate fibers generally require a certain degree of care in laundering, although they exhibit greater wrinkle recovery on drying than cellulosic fibers. Both acetate and triacetate fabrics do not shrink or lose their shape readily during laundering and can be laundered satisfactorily under mild laundry conditions. Both acetate and triacetate may be satisfactorily dry cleaned in Stoddard solvents, but certain other solvents must be avoided due to fiber solubility.

Acetate and triacetate dry quickly and may be tumble dried or drip dried. During tumble drying of these materials it is essential that the dryer be cool during the final tumbling.

Both acetate and triacetate can be safely ironed, but direct contact of the fabric with the iron is not recommended. Acetate must be ironed at a lower temperature than triacetate, and the maximum safe ironing temperature for these fabrics is 175°C and 200°C, respectively. The flammability characteristics of these fibers are similar to those of the cellulosic fibers, and they have LOIs of 18.

Acetate is used in dresses, blouses, foundation garments, lingerie, garment linings, some household furnishings, and certain specialty fabrics. Triacetate is used in sportswear, tricot fabrics, and in garments where pleats and pleat retention is important, as well as in certain specialty fabrics.

5. Protein Fibers

The protein fibers are formed by natural animal sources through condensation of α-amino acids to form repeating polyamide units with various substituents on the α-carbon atom. The sequence and type of amino acids making up the individual protein chains contribute to the overall properties of the resultant fiber. Two major classes of natural protein fibers exist and include keratin (hair or fur) and secreted (insect) fibers. In general, the keratin fibers are proteins highly crosslinked by disulfide bonds from cystine residues in the protein chain, whereas secreted fibers tend to have no crosslinks and a more limited array of less complex amino acids. The keratin fibers tend to have helical portions periodically within protein sequence, whereas the secreted fiber protein chains are arranged in a linear pleated sheet structure with hydrogen bonding between amide groups on adjacent protein chains. Keratin fibers are extremely complex in structure and include a cortical cell matrix surrounded by a cuticle sheath laid on the surface as overlapping scales. The cell matrix of some coarser hair fibers may contain a center cavity or medulla. The keratin fibers are round in cross section with an irregular crimp along the longitudinal fiber axis which results in a bulky texturized fiber. Secreted fibers are much less complex in morphology and often have irregular cross sections. Other fibrous protein materials include azlon fibers, spun from dissolved proteins, and a graft protein-acrylonitrile matrix fiber.

In general, protein fibers are fibers of moderate strength, resiliency, and elasticity. They have excellent moisture absorbency and transport characteristics. They do not build up static charge. While they have fair acid resistance, they are readily attacked by bases and oxidizing agents. They tend to yellow in sunlight due to oxidative attack. They are highly comfortable fibers under most environmental conditions and possess excellent aesthetic qualities.

$$\left[\begin{array}{c} \text{CHCNH} \\ | \\ \text{R} \end{array}\right]_n$$

PROTEIN

R	AMINO ACID	R	AMINO ACID
H –	GLYCINE	$HOCH_2$ –	SERINE
CH_3 –	ALANINE	$CH_3CH–$ $\quad OH$	THREONINE
$(CH_3)_2CH$ –	VALINE	$-CH_2SSCH_2$ –	CYSTINE
$(CH_3)_2CHCH_2$ –	LEUCINE	$CH_3SCH_2CH_2$ –	METHIONINE
$CH_3CH_2CH–$ $\quad CH_3$	ISOLEUCINE	NH $NH_2CNH(CH_2)_3$ –	ARGININE
PROLINE		HISTIDINE	
CH_2 – PHENYLALANINE		$NH_2(CH_2)_4$ – LYSINE	
HO—CH_2 – TYROSINE		$HOOC\ CH_2$ – ASPARTIC ACID	
CH_2 – TRYPTOPHAN		$HOOC(CH_2)_2$ – GLUTAMIC ACID	

WOOL

Wool is a natural highly crimped protein hair fiber derived from sheep. The fineness and the structure and properties of the wool will depend on the variety of sheep from which it was derived. Major varieties of wool come from Merino, Lincoln, Leicester, Sussex, Cheviot, and other breeds of sheep. Worsted wool fabrics are made from highly twisted yarns of long and finer wool fibers, whereas woolen fabrics are made from less twisted yarns of coarser wool fibers.

Structural Properties

Wool fibers are extremely complex, highly crosslinked keratin proteins made up of over 17 different amino acids. The amino acid content and sequence in wool varies with variety of wool. The average amino acid contents for the major varieties of wool are given in Table 5-1.

Table 5-1. Amino Acid Contents in Wool Keratins

Amino Acid	Content in Keratin (g/100 g Wool)
Glycine	5-7
Alanine	3-5
Valine	5-6
Leucine	7-9
Isoleucine	3-5
Proline	5-9
Phenylalanine	3-5
Tyrosine	4-7
Tryptophan	1-3
Serine	7-10
Threonine	6-7
Cystine	10-15
Methionine	0-1
Arginine	8-11
Histidine	2-4
Lysine	0-2
Aspartic acid	6-8
Glutamic acid	12-17

The wool protein chains are joined periodically through the disulfide crosslinked cystine, a diamino acid that is contained within two adjacent chains. About 40% of the protein chains spiral upon themselves and internally hydrogen bond to form an α-helix. Near the periodic cystine crosslinks or at points where proline and other amino acids with bulky groups occur along the chain, the close packing of chains is not possible and a less regular nonhelical structure is observed. The crosslinked protein structure packs and associates to form fibrils, which in turn make up the spindle shaped cortical cells which constitute the cortex or interior of the fiber. The cortex is made up of highly and less cross-linked <u>ortho</u> and <u>para</u> cortex positions. The cortex is surrounded by an outer sheath of scalelike layers or cuticle, which accounts for the scaled appearance running along the surface of the fiber (Figures 5-1 and 5-2).

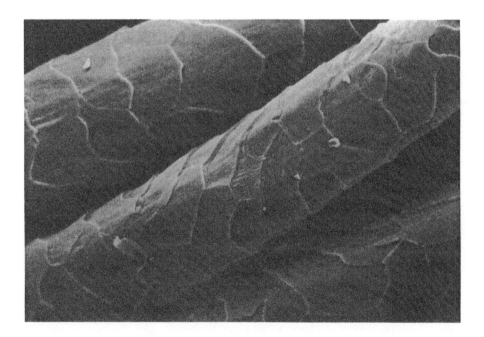

Figure 5-1. Wool. x900.

Physical Properties

Wool fibers possess low to moderate strength with tenacities of 1-2 g/d (9-18 g/tex) dry and 0.8-1.8 g/d (7-16 g/tex) wet. Elongations at break vary from 25% to 40% dry and 25% to 60% wet. At 2% extension, wool shows 99% recovery, and even at 20% extension a recovery as high as 65% is observed. Wool fibers have excellent resiliency and recover readily from deformation except under high humidities. The stiffness of wool varies according to the source and the diameter of the individual fibers. The moisture regain of wool is very high and varies between 13% and 18% under standard conditions. At 100% RH, the regain approaches 40%. Wool fibers have specific gravities of 1.28-1.32. Wool is insoluble in all solvents except those capable of breaking the disulfide crosslinks, but it does tend to swell in polar solvents. Wool is little affected by heat up to 150°C and is a good heat insulator due to its low heat conductivity and bulkiness, which permits air entrapment in wool textile structures. At moderate humidities, wool does not build up significant static charge.

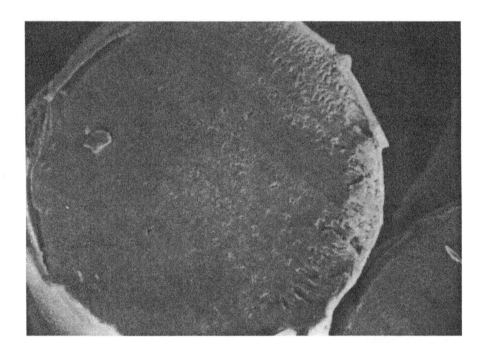

Figure 5-2. Wool cross-section. x2400.

Chemical Properties

Wool is resistant to attack by acids but is extremely vulnerable to attack by weak bases even at low dilutions. Wool is irreversibly damaged and colored by dilute oxidizing bleaches such as hypochlorite. Reducing agents cause reductive scission of disulfide bonds within the wool, eventually causing the wool to dissolve. Under controlled conditions, reducing agents can be used to partially reduce the wool and flat set or set permanent pleats in the wool. Unless chemically treated, wool is susceptible to attack by several species of moths able to dissolve and digest wool fibers. Wool is quite resistant to attack by other biological agents such as mildew. Wool is attacked by short wavelength (300-350 nm) ultraviolet light, causing slow degradation and yellowing. On heating, wool degrades and yellows above 150°C and chars at 300°C.

End-Use Properties

Wool varieties include Merino, Lincoln, Leicester, Sussex, Cheviot, Ramboullett, and Shetland, as well as many others. Wool is a fiber of high to moderate luster. Fabrics of wool possess a soft to moderate hand and exhibit good drapability. Wool fibers are highly absorbent and have excellent moisture transmission properties.

The low to moderate strength of wool fibers is compensated for by its good stretch and recovery properties. Wool is fairly abrasion resistant and does not tend to form pills due to its low strength. It resists wrinkling except under warm, moist conditions. Its crease retention is poor unless creases have been set using chemical reducing agents.

Wool is attacked by alkalies and chlorine bleaches and is progressively yellowed by the short ultraviolet wavelengths in sunlight. Wool dyes readily, and the dyed wools exhibit good colorfastness. Owing to its felting action, wool cannot be laundered in hot water with agitation, but it can be dry-cleaned or washed in warm water with a mild detergent if no agitation is used.

Due to its affinity for water, wool is slow drying. Wool may be ironed at 150°C or below without steaming. Wool is a self-extinguishing fiber and burns very slowly even in contact with a flame. It has an LOI of 25.

Wool is extensively used in textile applications where comfort and aesthetics are important. It is used in men's and women's apparel, outer wear and cold weather clothing, suits, blankets, felts, and carpeting. It is often used in blends with cellulosic and man-made fibers.

SILK

Silk is a natural protein fiber excreted by the moth larva <u>Bombyx</u> <u>mori</u>, better known as the common silkworm. Silk is a fine continuous monofilament fiber of high luster and strength and is highly valued as a prestige fiber. Because of its high cost, it finds very limited use in textiles. A minor amount of wild tussah silk is produced for specialty items. Attempts have been made to commercialize silk from spiders over the years, but all ventures have met with failure. Domestic and wild silks are essentially uncrosslinked and relatively simple in amino acid composition compared to the keratin fibers. The properties for silk listed here are for silk formed by <u>Bombyx</u> <u>mori</u> moth larvae.

Liquid silk protein is extruded from two glands in the head of the silkworm. The fibers emerge from a common exit tube or spinneret and harden into a single strand by a protein gum called sericin. The completed silk cocoons are soaked in hot water to loosen the sericin, and the silk filaments are unwound. After unwinding, the silk filaments are washed in warm detergent solutions to remove the sericin. The fibroin silk fibers are more simple in structure than keratin and are composed predominantly of glycine, alanine, tyrosine, and serine. The average range of composition for silk is given in Table 5-2.

With no cystine present in the fibroin protein, little crosslinking is observed between protein chains. The degree of polymerization of silk fibroin is uncertain, with DPs of 300 to 3000+ having been measured in different solvents. In the absence of crosslinks and with limited bulky side chains present in the amino acids, fibroin molecules align themselves parallel to each other and hydrogen bond to form a highly crystalline and oriented "pleated-sheet" or "beta" structure. Silk fibers are smooth surfaced and translucent with some irregularity in diameter along the fiber (Figure 5-3). The fibers are basically triangular in cross section with rounded corners.

Physical Properties

Silk fibers are strong with moderate degrees of recovery from deformation. Silk has a dry tenacity of 3-6 g/d (27-54 g/tex) and a wet tenacity of 2.5-5 g/d (23-45 g/tex). Silk exhibits a recovery of 90% from 2% elongation and of 30%-35% from 20% elongation. Silk fibers are moderately stiff and exhibit good to excellent resiliency and recovery from deformation, depending on temperature and humidity conditions. Silk has a specific gravity of 1.25-1.30 and a moisture regain of 11% under standard conditions. Silk is soluble in hydrogen bond breaking solvents such as aqueous lithium bromide, phosphoric acid, and cuprammonium solutions. It exhibits good heat insulating properties and is little affected by heat up to 150°C. Silk has moderate electrical resistivity and tends to build up static charges.

Table 5-2. Amino Acid Contents in Fibroin

Amino Acid	Content in Fibroin (g/100 g Fibroin)
Glycine	36-43
Alanine	29-35
Tyrosine	10-13
Serine	13-17
Valine	2-4
Leucine	0-1
Proline	0-1
Phenylalanine	1-2
Tryptophan	0-1
Threonine	1-2
Cystine	0
Methionine	0
Arginine	0-2
Histidine	0-1
Lysine	0-1
Aspartic acid	1-3
Glutamic acid	1-2

Chemical Properties

Silk is slowly attacked by acids but is damaged readily by basic solutions. Strong oxidizing agents such as hypochlorite rapidly discolor and dissolve silk, whereas reducing agents have little effect except under extreme conditions. Silk is resistant to attack by biological agents but yellows and loses strength rapidly in sunlight. Silk is often weighted with tin and other metal salts. These salts make silk even more sensitive to light-induced oxidative attack. Silk undergoes charring and oxidative decomposition when heated above 175°C in air over a prolonged period of time.

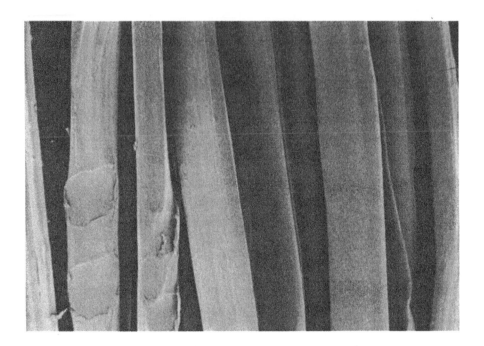

Figure 5-3. Silk. x900.

End-Use Properties

Silk possesses a combined set of aesthetic properties that make it useful for high-fashion luxury textile goods. Silk has a high luster and is translucent. Silk fabrics have pleasing appearance and drapability, and a characteristically pleasing crisp hand. Silk is highly moisture absorbent and has good to excellent resistance to wrinkling. It is a moderately strong fiber with moderate recovery properties. It exhibits fair abrasion resistance and good resistance to pilling. Silk is sensitive to chlorine bleaches and to alkalies and is easily damaged by sunlight. The fiber may be dyed with a wide variety of dyes to give dyed fibers with high color-fastness. Silk may be laundered under mild, nonalkaline conditions and dry cleans readily. Because of its high affinity for water, it dries slowly but may be dried or ironed safely up to 150°C. Silk burns slowly and self-extinguishes when removed from a flame. Silk is used extensively in luxury fabrics and apparel and home furnishings, and in accessories such as scarfs.

OTHER NATURAL AND REGENERATED PROTEIN FIBERS

The other major hair (keratin) fibers include mohair, cashmere, llama, alpaca, and vicuna, as well as many others. Regenerated Azlon fibers are derived from soluble proteins that can be spun into fibers, insolubilized, and regenerated. Soluble proteins also may be grafted to form a copolymer, dissolved, and then spun into fibers.

Mohair

Mohair is a very resilient hair fiber obtained from the angora goat. The two primary classifications for mohair are the finer kid mohair and the coarser adult mohair. In many respects, mohair resembles wool in structure and possesses properties including the characteristic scale structure of the fiber. The average length of mohair fibers is longer than wool, with 4-12 inch fiber lengths being typical. It is a much stronger fiber than wool, but its other tensile properties resemble wool. Mohair is remarkably resistant to wear and is used in applications where such durability is essential. Mohair possesses a beautiful natural luster and is used in blends to provide luster. Mohair fibers also provide a characteristic, resilient, and slightly scratchy hand in even small quantities blended with other fibers.

Cashmere

Cashmere is the fine, soft inner coat of down obtained from the cashmere goat found on the inner plateaus of Asia. In many ways the properties of cashmere resemble those of wool, but cashmere fibers are extremely fine and soft compared to wool. Cashmere is used in luxury applications where a soft, warm, fine fiber with beautiful drape is desired.

Llama, Alpaca, and Vicuna

These fibers come from a group of related animals found in South America. They are fine fibers that are white to tan and brown in color. They are longer than most wool fibers and generally stronger, with a finer scale structure. They are generally used only in the expensive luxury items of textiles and apparel.

Regenerated Protein Fibers

Azlon is the generic name given to manufactured fibers composed of a regenerated natural protein. Azlon is produced by dissolving proteins like casein from milk, soya bean protein, and zein from corn in dilute alkali and forcing these solutions through a spinneret into an acid-formaldehyde coagulating bath. Many of the properties of these fibers resemble the natural protein fibers, but they suffer from low dry and wet strength and sensitivity to alkalies. Although no longer produced in the U.S., azlon fibers are produced in Europe and used in blends with natural and man-made fibers.

Protein-Polyacrylonitrile Graft Copolymer

A fiber consisting of a copolymer of casein protein (25%-60%) grafted with 40%-75% acrylic monomers, of which at least half is acrylonitrile, has been developed in Japan under the tradename Chinon. The casein dissolved in aqueous zinc chloride and grafted with acrylonitrile is wet or dry spun into fibers. The fiber has a tenacity of 3.5-5 g/d (32-45 g/tex) dry and 3-4.5 g/d (27-40 g/tex) wet and an elongation at break of 15%-25% wet or dry. It recovers 70% from 5% elongation. The fiber has a moisture regain of 4.5%-5.5% and a specific gravity of 1.22. It dyes readily with acid dyes, but basic and reactive dyes can be used also. The fiber is marketed as a substitute for silk.

6. Polyamide Fibers

The polyamide fibers include the nylons and the aramid fibers. Both fiber types are formed from polymers of long-chain polyamides. In nylon fibers less than 85% of the polyamide units are attached directly to two aromatic rings, whereas in aramid fibers more than 85% of the amide groups are directly attached to aromatic rings. The nylons generally are tough, strong, durable fibers useful in a wide range of textile applications. The fully aromatic aramid fibers have high temperature resistance, exceptionally high strength, and dimensional stability. The number of carbon atoms in each monomer or comonomer unit is commonly designated for the nylons. Therefore the nylon with six carbon atoms in the repeating unit would be nylon 6 and the nylon with six carbons in each of the monomer units would be nylon 6,6.

NYLON 6 AND 6,6

Nylon 6 and 6,6 are very similar in properties and structure and therefore will be described together. The major structural difference is due to the placement of the amide groups in a continuous head-to-head arrangement in nylon 6, whereas in nylon 6,6 the amide groups reverse direction each time in a head-to-tail arrangement due to the differences in the monomers and polymerization techniques used:

$$\left[(CH_2)_5 \overset{\overset{\displaystyle O}{\|}}{C} NH \right]_n$$

NYLON 6

$$\left[\overset{\overset{\displaystyle O}{\|}}{C} (CH_2)_4 \overset{\overset{\displaystyle O}{\|}}{C} NH(CH_2)_6 NH \right]_n$$

NYLON 6,6

Nylon 6,6 was developed in the United States, whereas nylon 6 was developed in Europe and more recently Japan. These nylon polymers form strong, tough, and durable fibers useful in a wide variety of textile applications. The major differences in the fibers are that nylon 6,6 dyes lighter, has a higher melting point, and a slightly harsher hand than nylon 6.

Structural Properties

Nylon 6 is produced by ring-opening chain growth polymerization of caprolactam in the presence of water vapor and an acid catalyst at the melt. After removal of water and acid, the nylon 6 is melt spun at 250°C-260°C into fibers. Nylon 6,6 is prepared by step growth polymerization of hexamethylene diamine and adipic acid. After drying, the nylon 6,6 is melt spun at 280°C-290°C into fibers. Both nylon 6 and 6,6 are drawn to mechanically orient the fibers following spinning.

The degree of polymerization (DP) of nylon 6 and 6,6 molecules varies from 100 to 250 units. The polyamide molecular chains lay parallel to one another in a "pleated sheet" structure similar to silk protein with strong hydrogen bonding between amide linkages on adjacent molecular chains. The degree of crystallinity of the nylon will depend on the degree of orientation given to the fiber during drawing. Nylon fibers are usually rodlike with a smooth surface or trilobal in cross section (see Figures 6-1 and 6-2). Multilobal (star) cross sections and other complex cross sections are also found. Side-by-side bicomponent fibers with a round cross section are also formed. Following orientation and heating, these bicomponent fibers form small helical crimps along the fiber to give a texturized fiber useful in many applications including women's hosiery. Cantrece nylon is such a fiber. Also, sheath-core nylons with differing melting characteristics are formed for use as self-binding fibers for nonwovens.

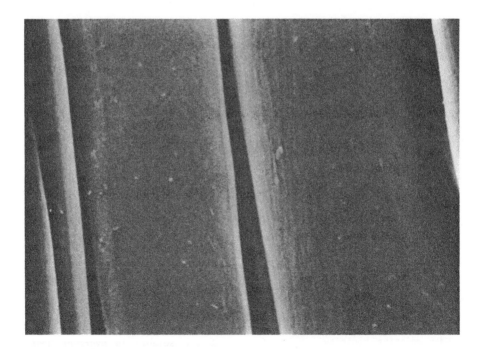

Figure 6-1. Nylon 6,6, round. x1900.

Figure 6-2. Nylon 6,6, trilobal. x1200.

Physical Properties

Nylon 6 and 6,6 fibers are strong, with a dry tenacity of 4-9 g/d (36-81 g/tex) and a wet tenacity of 2.5-8 g/d (23-72 g/tex). These nylons have elongations at break of 15%-50% dry, which increase somewhat on wetting. Recovery from stretch deformation is very good, with 99% recovery from elongations up to 10%. The nylons are stiff fibers with excellent resiliency and recovery from bending deformation. They are of low density, with a specific gravity of 1.14. They are moderately hydrophilic with a moisture regain of 4%-5% under standard conditions and a regain of 9% at 100% relative humidity. Nylon 6 and 6,6 are soluble in hydrogen bond breaking solvents such as phenols, 90% formic acid, and benzyl alcohol. They have moderate heat conductivity properties and are unaffected by heating below 150°C. The nylons have a high resistivity and readily build up static charge.

Chemical Properties

The nylons are fairly resistant to chemical attack. They are attacked by acids, bases, and reducing and oxidizing agents only under extreme conditions. They are unaffected by biological agents, but at elevated temperatures or in the presence of sunlight they will undergo oxidative degradation with yellowing and loss of strength.

End-Use Properties

Nylon 6 and 6,6 are marketed under several trade names. In fact, nylon was originally used as a trade name for duPont polyamide fiber, but as a result of common usage the term nylon came to be the generic term for these polyamide fibers. Common trade names for these nylons include Anso, Antron, Cadon, Cantrece, Cordura, Caprolan, duPont nylon, and Enkalure. Nylon 6 and 6,6 are extremely strong fibers with excellent recovery and resiliency properties. Nylon fabrics have a fair hand, with nylon 6 having a somewhat softer hand than nylon 6,6. Nylon fabrics have high luster unless delustered. They have moderate to excellent draping properties depending on the denier of the fiber. Unless bleached and dyed, nylons have a slightly yellow color. The nylons are wrinkle resistant and have good crease resistance if heat set. The fiber is tough and has good abrasion resistance. In fact, nylon is reported to abrade other fibers in a fiber blend. The fiber has good resistance to household chemicals but exhibits poor resistance to attack by sunlight unless treated with antioxidants. The fibers have excellent dyeability with excellent colorfastness properties. Nylon 6 is somewhat deeper dyeing than nylon 6,6.

The fiber has good laundering and dry cleaning properties but tends to scavenge dyes bleeding from other fibers. The fibers have moderate moisture uptakes and dry readily at temperatures up to 150°C. Nylon 6 can be safely ironed up to 140°C, whereas nylon 6,6 can be safely ironed up to 180°C. The nylons are less flammable than cellulosics, with an LOI of 20. They melt and drip and tend to self-extinguish on burning. Nylons are extensively used in hosiery, lingerie, underwear, sweaters, and other knitted goods. They have been used extensively in light and sheer apparel articles such as windbreakers. Nylons are used in home furnishings and industrial applications including carpets, upholstery, tie cord, parachutes, sails, ropes, thread, and outdoor wear.

ARAMID FIBERS

The aramid polyamide fibers are formed from a long chain of synthetic polyamides in which at least 85% of the amide linkages are attached to aromatic rings. These essentially fully aromatic polyamides are characteristically high melting and have excellent property retention at high temperatures and excellent durability. They are unaffected by moisture and most chemicals and are inherently flame retardant. The fibers have high strength and can be used in a number of unique high-strength applications.

Structural Properties

Aramids are formed through step growth polymerization of aromatic diacid chlorides with aromatic diamines in a polar aprotic solvent such as N,N-dimethylformamide (DMF) to a DP of 100-250. The meta- and para-substituted benzene dicarboxylic acid chlorides and diamines are characteristically used for aramid fibers presently in production, but other fully aromatic ring systems are possible future sources of aramid polymers for fibers:

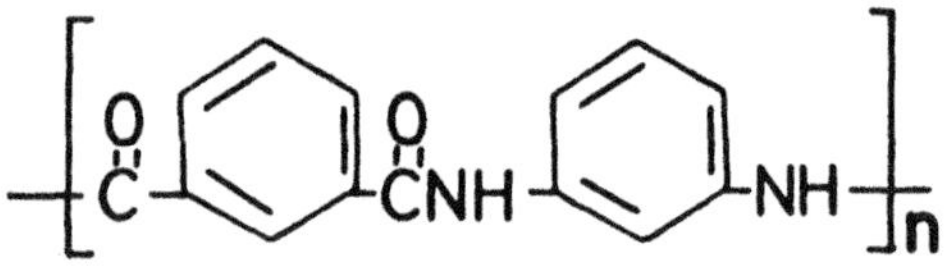

NOMEX

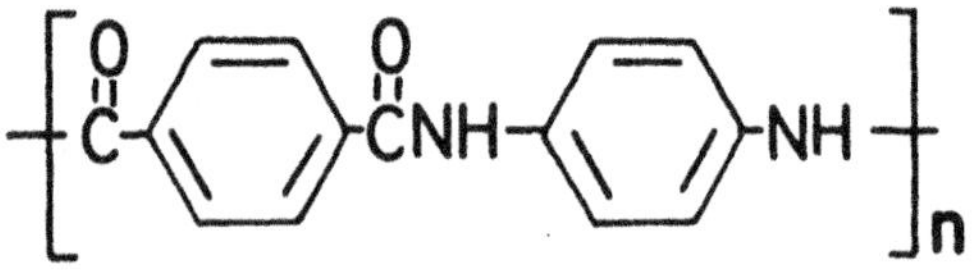

KEVLAR

ARAMID

The resultant aramid polymers are spun in suspension through a spinneret into hot air (dry spinning) or a coagulating of both (wet spinning) to form the fibers, followed by fiber stretching and orientation. The aromatic units along aramid polyamide chains confer stiffness to the polymeric chains and limit their flexibility and mobility. Hydrogen bonding between amide groups on adjacent chains and extremely strong van der Waals interactions between aromatic rings planar to adjacent aromatic rings provide a tightly-packed, strongly held molecular structure and account for the strength and thermal stability of the aramids. The aramids are usually spun in round or dumbbell cross section.

Physical Properties

The aramids are the strongest of the man-made fibers, with the strength depending on the polymer structure, spinning method used, and the degree of orientation of the fiber. Dry tenacities of 4-22 g/d (36-198 g/tex) are found for the aramids, and wet tenacities of 3-18 g/d (27-162 g/tex) have been recorded. The elongation at break of the fibers decreases with increasing tenacity over a range of 3%-30% dry and wet. Recovery from low elongations of 5% or less is 98%-100%. The fibers are stiff, with excellent resiliency and recovery from bending deformation. The specific gravities of these fibers vary from 1.38 to 1.44. The aramids have a moisture regain comparable to the other polyamide fibers and are in the 3.5%-7% range. The fibers are swollen and dissolved in polar aprotic solvents or strong acids. The aramids have high heat and electrical resistivities, have excellent insulative capabilities, and are unaffected by heat up to 250°C.

Chemical Properties

The aramids are extremely resistant to chemical or biological attack. Only under extreme conditions and at elevated temperatures will concentrated acids or strong oxidizing agents attack aramids. Aramid fibers undergo initial oxidative attack in sunlight, causing discoloration and slight strength loss, but further exposure has little additional effect. The aramids also act as effective screens to high-energy nuclear radiation due to their ability to trap and stabilize radical and ionic species induced by radiation. The aramids undergo oxidative degradation on prolonged heating above 370°C.

End-Use Properties

Common trade names for aramid fibers include Nomex and Kevlar (duPont). Aramid fibers are extremely strong and heat resistant. Fabrics from the aramids have a high luster with a fair hand and adequate draping properties. The fibers are light yellow unless bleached and exhibit moderate moisture absorption characteristics. The fibers recover readily from stretching and bending deformation and are extremely abrasion resistant. They do tend to pill due to the high strength of the fiber. The aramids are extremely resistant to attack by household chemicals and exhibit good resistance to sunlight. The fibers are difficult to dye except by special dyeing techniques with disperse dyes. Dyed aramids are reasonably colorfast. They have good launderability and dry-cleanability and are moderately easy to dry. The fibers are extremely heat stable and can be ironed up to 300°C. The fibers are of low flammability and self-extinguish on removal from a flame. They have an LOI of 30. Aramid fibers are used in fabric applications where high strength and low flammability are important, including protective clothing, electrical insulation, filtration fabrics, specialized military applications, and tire cord.

OTHER POLYAMIDES

Several other polyamides have been introduced for use as fibers in specialty applications where certain combinations of properties are desired. The major specialty nylons include Qiana, nylon 4, nylon 11, nylon 6,10, and biconstituent nylon-polyester. Their particular characteristics are outlined below.

Qiana

Qiana is a trade name (duPont) for the luxury nylon fiber formed through step growth polymerization of <u>trans,trans</u>-di(4-aminocyclohexyl) methane and a dibasic acid having 8-12 carbon atoms:

$$\left[-NH \left\langle \begin{array}{c} CH_2 \\ \end{array} \right\rangle -NH\overset{O}{\underset{\parallel}{C}}(CH_2)_X\overset{O}{\underset{\parallel}{C}} \right]_n$$

QIANA

Qiana resembles nylon 6 and 6,6 in many of its properties but also has a unique silklike texture, a lustrous appearance, excellent form retention, and an overall superior performance level which makes it useful in prestige textile products and in the sophisticated fashion apparel market.

Qiana has a tenacity of 3-3.5 g/d (27-32 g/tex) and an elongation at break of 20%-30%. It is reported to recover from strain better than nylon 6,6 or polyester. It has a specific gravity of 1.03 and a moisture regain of 2%-2.5%, which are much lower values than nylon 6 or 6,6. Qiana melts at 275°C and is stable up to 185°C. The cross section of the fiber is tri-lobal (Figure 6-3). It shows good chemical stability and is readily dye-able to fast colors. The excellent aesthetic qualities and drapability of Qiana have contributed greatly to its success.

Figure 6-3. Qiana. x900.

Nylon 4

Nylon 4 is produced by polymerization of pyrrolidone using carbon dioxide catalyst. Nylon 4 is melt spun (melting point 273°C) to give a fiber of about 60% crystallinity. The fiber is of moderate dry strength (4.5 g/d) and slightly higher wet strength, with a higher specific gravity (1.25) than nylon 6 and 6,6. The fiber has excellent elongation and recovery properties and a water regain comparable to cotton (8% at 21°C and 65% RH). Nylon 4 is easily laundered and can be dyed easily to give colorfast shades but has limited wrinkle recovery properties and is sensitive to hypochlorite bleaches. Although a major attempt was made to commercialize nylon 4, market conditions limited the penetration of this new fiber type and production was discontinued. Nylon 4 is no longer being produced.

Nylon 11

Nylon 11 is produced by self-condensation of 11-aminoundecanoic acid in Europe and marketed under the name Rilsanite. It is melt spun (mp 189°C) into fiber and possesses essentially the same properties as nylon 6 and 6,6. It has high strength (5-7.5 g/d), low specific gravity (1.04), and low water regain (1.2%). Nylon 11 has excellent electrical properties and is used in electrical products as well as in brush bristles, tire cord, lingerie, and hose.

Nylon 6,10

Nylon 6,10 is produced by condensation polymerization of hexamethylene diamine and sebacic acid and is melt spun (mp 216°C) into fibers. It resembles nylon 6 and 6,6 in many ways but has a lower moisture regain (2.6%). It is primarily used in brush bristles.

Biconstituent Nylon-Polyester

Biconstituent fiber of nylon 6 with polyester microfibrils dispersed throughout the fiber matrix has been marketed under the trade name Source (Allied). The fiber is reported to have unique optical and dyeing properties and a higher strength and lower regain than nylon 6,6 and is used primarily in carpets. A sheath-core bicomponent fiber containing a nylon 6 sheath and a polyester core has been reported also. It is said to have properties that are intermediate between both fibers.

7. Polyester Fibers

Polyesters are those fibers containing at least 85% of a polymeric ester of a substituted aromatic carboxylic acid including but not restricted to terephthalic acid and p-hydroxybenzoic acid. The major polyester in commerce is polyethylene terephthalate, an ester formed by step growth polymerization of terephthalic acid and the diol ethylene glycol. Poly-1,4-cyclohexylenedimethylene terephthalate is the polyester of more limited usage and is formed through the step growth polymerization of terephthalic acid with the more complex diol 1,4-cyclohexylenedimethanol. A third polyester fiber is actually the polyester ether poly(p-ethyleneoxybenzoate) and is formed through step growth polymerization of p-hydroxybenzoic acid with ethylene glycol; it is no longer in commercial production, however. The polyester fibers all have similar properties, are highly resilient and resistant to wrinkling, possess high durabilily and dimensional stability, and are resistant to chemical and environmental attack.

POLYETHYLENE TEREPHTHALATE

Polyethylene terephthalate polyester is the leading man-made fiber in production volume and owes its popularity to its versatility alone or as a blended fiber in textile structures. When the term "polyester" is used, it refers to this generic type. It is used extensively in woven and knitted apparel, home furnishings, and industrial applications. Modification of the molecular structure of the fiber through texturizing and or chemical finishing extends its usefulness in various applications. Polyester is expected to surpass cotton as the major commodity fiber in the future.

Structural Properties

Polyethylene terephthalate is formed through step growth polymerization of terephthalic acid or dimethyl terephthalate with ethylene glycol at 250°-300°C in the presence of a catalyst to a DP of 100-250:

$$\left[\,\overset{O}{\underset{\parallel}{C}}\,\text{—}\!\!\left\langle\right\rangle\!\!\text{—}\,\overset{O}{\underset{\parallel}{C}}\text{OCH}_2\text{CH}_2\text{O}\,\right]_n$$

POLYETHYLENE TEREPHTHALATE
POLYESTER

The resultant polymer is isolated by cooling and solidification and dried. Polyester fibers are melt spun from the copolymer at 250°-300°C, followed by fiber orientation and stretching. The polyester molecular chains are fairly stiff and rigid due to the presence of periodic phenylene groups along the chain. The polyester molecules within the fiber tend to pack tightly and are held together by van der Waals forces. The polyesters are highly crystalline unless comonomers are introduced to disrupt the regularity of the molecular chains. Polyester fibers are usually smooth and rodlike with round or trilobal cross sections (Figure 7-1).

Figure 7-1. Polyester. x1000.

Physical Properties

Polyester from polyethylene terephthalate is an extremely strong fiber with a tenacity of 3-9 g/d (27-81 g/tex). The elongation at break of the fiber varies from 15% to 50% depending on the degree of orientation and nature of crystalline structure within the fiber. The fiber shows moderate (80%-95%) recovery from low elongations (2%-10%). The fiber is relatively stiff and possesses excellent resiliency and recovery from bending deforma-tion. The fiber has a specific gravity of 1.38. The fiber is quite hydro-phobic, with a moisture regain of 0.1%-0.4% under standard conditions and 1.0% at 21°C and 100% RH. It is swollen or dissolved by phenols, chloro-acetic acid, or certain chlorinated hydrocarbons at elevated temperatures. The fiber exhibits moderate heat conductivity and has high resistivity, leading to extensive static charge buildup. On heating, the fiber softens in the 210°-250°C range with fiber shrinkage and melts at 250°-255°C.

Chemical Properties

Polyethylene terephthalate polyester is highly resistant to chemical attack by acid, bases, oxidizing, or reducing agents and is only attacked by hot concentrated acids and bases. The fiber is not attacked by biological agents. On exposure to sunlight, the fiber slowly undergoes oxidative attack without color change with an accompanying slow loss in strength. The fiber melts at about 250°C with only limited decomposition.

End-Use Properties

Common trade names for polyethylene terephthalate polyester are Monsanto, Dacron, Encron, Kodel IV, Polar Guard, Trevira, Fortrel, and Vycron. This polyester is an inexpensive fiber with a unique set of desirable properties which have made it useful in a wide range of end-use applications. It has adequate aesthetic properties and bright translucent appearance unless a delusterant has been added to the fiber. The hand of the fiber is somewhat stiff unless the fiber has been texturized or modified, and fabrics from polyester exhibit moderate draping qualities.

The fiber is hydrophobic and nonabsorbent without chemical modification. This lack of absorbency limits the comfort of polyester fabrics. Polyester possesses good strength and durability characteristics but exhibits moderate to poor recovery from stretching. It has excellent wrinkle resistance and recovery from wrinkling and bending deformation. Polyester shows only fair pilling and snag resistance in most textile constructions. Heat setting of polyester fabrics provides dimensional stabilization to the structure, and creases can be set in polyester by heat due to its thermoplastic nature. It has excellent resistance to most household chemicals and is resistant to sunlight-induced oxidative damage, particularly behind window glass.

Due to its hydrophobicity and high crystallinity, polyester is difficult to dye, and special dyes and dyeing techniques must be used. When dyed, polyester generally exhibits excellent fastness properties. Polyester has good laundering and dry-cleaning characteristics but is reported to retain oily soil unless treated with appropriate soil release agents. Owing to its low regain, polyester dries readily and can be safely ironed or dried at temperatures up to 160°C. It is a moderately flammable fiber that burns on contact with a flame but melts and drips and shrinks away from the flame. It has a LOI 21. Polyester is used extensively as a staple and filament fiber in apparel and as a staple fiber in blends with cellulosic fibers for apparel of all types. Polyester is also used extensively in home furnishings and as fiberfill for pillows, sleeping bags, etc. Polyester is used in threads, rope, and tire cord and in sails and nets as well as other industrial fabrics.

POLY-1,4-CYCLOHEXYLENEDIMETHYLENE TEREPHTHALATE

Poly-1,4-cyclohexylenedimethylene terephthalate resembles polyethylene terephthalate in most properties:

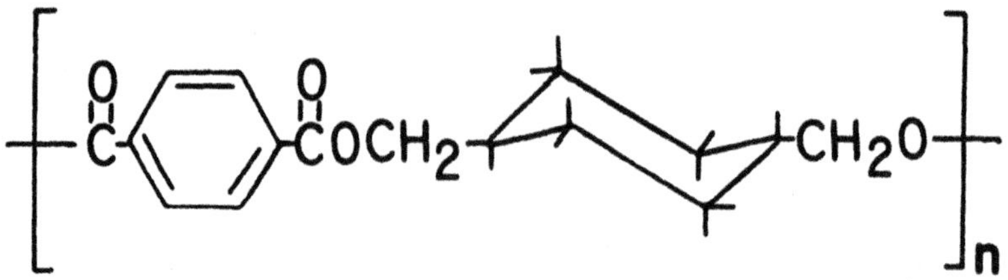

POLY-I,4-CYCLOHEXYLENEDIMETHYLENE TEREPHTHALATE   POLYESTER

The cyclohexylene group within this fiber provides additional rigidity to the molecular chains, but the packing of adjacent polymer chains may be more difficult due to the complex structure (Figure 7-2). As a result, the fiber has a lower tenacity than polyethylene terephthalate. It has a tenacity of 2.5-3 g/d (22-27 g/tex) and exhibits lower elongations than polyethylene terephthalate. It has a lower specific gravity (1.22-1.23) than polyethylene terephthalate. The fiber melts at 290°-295°C and is attacked and shrunk by trichloroethylene and methylene chloride. The fiber has good chemical resistance. The major trade name for this polyester is Kodel II. The fiber is somewhat superior to polyethylene terephthalate in certain end-use properties including better recovery from stretch and better resistance to pilling. The fiber has superior resiliency and is particularly suited for use in blend with cellulosics and wool, as a carpet fiber, and as fiberfill.

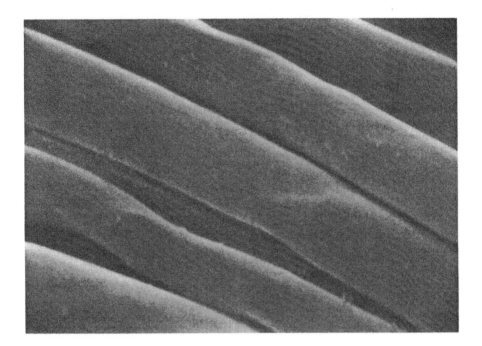

Figure 7-2. Poly-1,4-cyclohexylenedimethylene terephthalate polyester. x1300.

OTHER POLYESTERS

Poly-p-ethyleneoxybenzoate

This fiber was introduced by a Japanese firm as A-Tell. The properties of the fiber are very similar to the polyesters containing terephthalate units in most respects including tensile properties, specific gravity, melting point, chemical resistance, and sunlight resistance. As a result, the FTC expanded the polyester category to include the polymeric p-ethyleneoxybenzoate esters. The fiber is produced by reaction of ethylene oxide with p-hydroxybenzoic acid, followed by melt spinning.

The fiber has a tenacity of 4-5.5 g/d (36-50 g/tex) and an elongation at break of 15%-30% with nearly complete recovery from low elongation. Its specific gravity of 1.34 and moisture regain of 0.4%-5.0% are nearly the same as that of polyethylene terephthalate polyester. The fiber melts at 224°C and softens at about 200°C. The fiber is reported to be even more resistant to attack by acids and bases than terephthalate-based polyesters. The properties of this fiber were not sufficiently different from other polyesters to achieve reasonable market penetration, and the fiber has been discontinued.

Modified Terephthalate Polyesters

Poor dyeability and the moderate flammability of polyester have resulted in formulation of modified terephthalate esters to improve the dyeability and the flame retardant properties of the fiber. Introduction of amino or sulfonic acid groups on the benzene ring of terephthalate leads to fibers that are more dyeable with cationic or acid dyes. Bromine, other halogen, or phosphonate groups substituted within the structure provide flame retardant characteristics to the fiber, especially if antimony oxide is present in the fiber matrix. Appropriate nonterephthalate comonomers replacing some of the terephthalate groups introduced into the polyester can improve dyeability and/or flammability of the fibers.

8. Acrylic Fibers

The acrylic fibers include acrylic, modacrylic, and other vinyl fibers containing cyanide groups as side chains. Among the major acrylic fibers used in commerce, acrylonitrile is the comonomer containing a cyanide group. Acrylic fibers are formed from copolymers containing greater than 85% acrylonitrile monomer units, whereas modacrylic fibers contain 35%-85% acrylonitrile units. Lastrile fibers contain 10%-50% acrylonitrile units copolymerized with vinylidene chloride monomer, whereas nytril fibers are made from copolymers containing at least 85% vinylidene dinitrile units. In general, these fibers possess a warm bulky hand, good resiliency and wrinkle resistance, and overall favorable aesthetic properties.

ACRYLIC

Acrylic fibers are formed from wet or dry spinning of copolymers containing at least 85% acrylonitrile units. After texturizing, acrylic fibers have a light bulky wool-like hand and overal wool-like aesthetics. The fibers are resilient and possess excellent acid resistance and sunlight resistance. Acrylics have been used extensively in applications formerly reserved for wool or other keratin fibers.

Structural Properties

Acrylic fibers are made up of copolymers containing at least 85% acrylonitrile units in combination with one or more comonomers including methyl methacrylate, vinyl acetate, or vinyl pyridine:

$$\left[-(CH_2CH)_x(CH_2\underset{R'}{\overset{R}{C}})_y- \right]_n$$

$$CN$$

$$x > 85\%, \; y < 15\%$$

$$R = H, -CH_3 \quad R' = -\overset{O}{\overset{\|}{C}}OCH_3, -O\overset{O}{\overset{\|}{C}}CH_3, \; \text{(pyridine ring)} \; , etc.$$

ACRYLIC

The copolymer is formed through free radical emulsion polymerization. After precipitation the copolymer is dried and dissolved in an appropriate organic solvent and wet or dry spun. The degree of polymerization of the copolymers used for fiber formation varies from 150 to 200 units. Pure polyacrylonitrile will form satisfactory fibers. Owing to the extensive tight packing of adjacent molecular chains and the high crystallinity of the fiber, comonomers must be introduced to lower the regularity and crystallinity of the polymer chains to make the fiber more dyeable. Extensive hydrogen bonding occurs between α-hydrogens and the electronegative nitrile groups on adjacent polymer chains, and strong van der Waals interactions further contribute to the packing of the acrylic chains. The periodic comonomer units interfere with this packing and therefore decrease the overall crystallinity of acrylic fibers. Acrylic fibers are usually smooth with round or dog-bone cross sections (Figure 8-1). Many bicomponent acrylic fibers are produced in order to provide a bulky texturized structure on drawing.

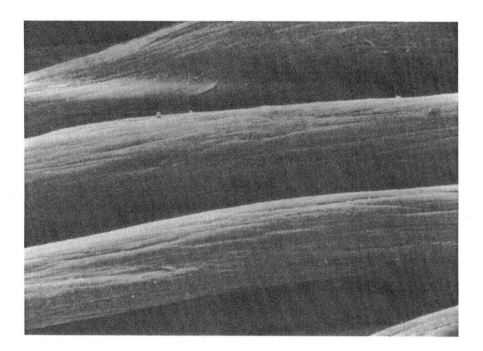

Figure 8-1. Acrylic. x1300.

Physical Properties

Acrylic fibers are fibers of moderate strength and elongations at break. The tenacity of acrylic fibers varies from 2 to 4 g/d (18-36 g/tex). On wetting, the tenacity drops to 1.5-3 g/d (13-27 g/tex). The elongation at break varies from 20% to 50% for the various acrylic fibers. At 2% elongation the recovery of the fiber is 99%; however, at 5% elongation the recovery is only 50%-95%. The fiber is moderately stiff and has excellent resiliency and recovery from bending deformation. The fibers have low specific gravities of 1.16-1.18 and low moisture regains of 1.0%-

2.5% under standard temperature and humidity conditions. The fiber is soluble in polar aprotic solvents such as N,N-dimethylformamide. The fiber exhibits good heat and electrical insulation properties. Acrylic fibers do build up moderate static charge and soften at 190°-250°C.

Chemical Properties

Acrylic fibers exhibit good chemical resistance. The fibers are only attacked by concentrated acids and are slowly attacked and hydrolyzed by weak bases. Acrylics are unaffected by oxidizing and reducing agents except for hypochlorite solutions at elevated temperatures. Acrylic fibers are unaffected by biological agents and sunlight. On heating above 200°C, acrylic fibers soften and undergo oxidative attack by a complex mechanism with formation of condensed unsaturated chromophoric (colored) groups in the fiber.

End-Use Properties

Common trade names for acrylic fibers include Acrilan, Creslan, Orlon, Sayelle, and Zefran. Since the acrylic fibers are usually texturized, they have a bulky wool-like hand and possess a moderate degree of luster. The fibers are of moderate strength but exhibit poor recovery from elongation. They possess fair abrasion and pilling resistance. The fiber has good wrinkle resistance and crease retention if properly heat set. The fiber possesses good resistance to household chemicals and sunlight and is moderately resistant to heat-induced oxidation and discoloration. The fiber undergoes laundering and dry-cleaning very well and may be dried readily due to its low affinity for moisture. The fiber may be ironed safely up to 150°C. Owing to the introduction of comonomer, acrylic fibers are generally dyeable and give fast colors with a wide range of dyes including acid, basic, or disperse dyes. The comonomer present determines the type of dye(s) that may be effectively used. The acrylic fibers are moderately flammable with a LOI of 18. The fibers burn with melting and continue to burn on withdrawal from the flame. On extinguishing, the fiber leaves a hard black bead. Acrylic fibers have found a wide spectrum of use where soft, warm, wool-like characteristics are desired. Such uses would include sweaters, skirts, dresses, suits, outerwear and knitted wear, blankets, socks, carpets, drapes, upholstery, and pile fabrics.

MODACRYLIC

Modacrylic fibers are formed from copolymers consisting of 35%-85% acrylonitrile and a suitable vinyl comonomer or comonomers such as vinyl chloride, vinylidene chloride, vinyl acetate, vinyl pyrollidone, or methyl acrylate. The modacrylics generally resemble acrylics and have a warm pleasing hand and good drapability, resiliency and wrinkle resistance. They are more heat sensitive but more flame resistant than acrylics and have generally been used in specialty applications. Modacrylic fiber exhibits a more thermoplastic character than the related acrylic fibers.

Structural Properties

Modacrylic fibers are wet or dry spun from copolymers of acrylonitrile (35%-85%) and an appropriate comonomer or comonomers:

$$\left[(CH_2CH)_{\overline{x}}(CH_2\underset{R'}{\overset{R}{C}})_y \right]_n$$

$$\text{CN}$$

$$x > 35\%, < 85\% \quad y > 15\%, < 65\%$$

$$R = H; CH_3; Cl \quad R' = -Cl, -O\overset{O}{\overset{\|}{C}}CH_3, -\overset{O}{\overset{\|}{C}}OCH_3, -N\langle\ \rangle$$

MODACRYLIC

The copolymer is formed through free radical chain growth emulsion or solution polymerization to a DP of 150-500. The copolymer is isolated and dissolved in acetone or a similar low-boiling-point solvent and wet or dry spun to a fiber of round, dog-bone, crescent, or polylobal cross section (Figure 8-2). The oriented fiber possesses a low crystallinity due to the irregularity and heterogeneity of the copolymer structure. Limited hydrogen bonding and van der Waals interactions are possible due to the limited regularity of adjacent polymer chains.

Figure 8-2. Modacrylic. x1000.

Physical Properties

The modacrylic fiber is of moderate strength with a dry tenacity of 1.5-3 g/d (14-27 g/tex) and a slightly reduced wet tenacity of 1-2.5 g/d (9-23 g/tex). The fiber has a high elongation at break of 25%-45% and excellent recovery (95%-100%) from low degrees of stretching (<5%). The fibers possess excellent resiliency and moderate stiffness and have specific gravities of 1.30-1.37. Modacrylic fibers exhibit a wide range of moisture regains, from 0.4% to 4%, depending on the nature and composition of comonomers making up the copolymer. Modacrylic fibers are soluble in ketone solvents such as acetone and in aprotic polar solvents including N,N-dimethylformamide. The fibers are good heat and electrical insulators but tend to build up static charge. The fibers soften in the 135°-160° range with accompanying heat shrinkage.

Chemical Properties

The modacrylic fibers exhibit excellent resistance to chemical agents. They exhibit good stability to light and biological agents. Modacrylic fibers melt at 190°-210°C with slight decomposition. SEF modacrylic fibers possess somewhat greater heat stability than the other modacrylics.

End-Use Properties

Modacrylic fiber trade names include Elura, SEF, Verel, and Zefran. Production of Dynel modacrylic fibers was discontinued in 1975. Modacrylics possess warm wool-like aesthetics and a generally bright luster. The fiber exhibits fair pilling and abrasion resistance. It has good wrinkle resistance and crease retention if the fiber has been properly heat set. The fiber has moderate strength and good recovery from low elongations. Modacrylics are resistant to attack by household chemicals and have excellent sunlight resistance. The laundry and dry-cleaning characteristics of the fiber are good if the temperature is held at moderate levels. The fibers are more difficult to dye with acid and/or basic dyes than acrylics. They have better affinity for disperse dyes and give dyeings of good colorfastness. Owing to the low moisture regain of modacrylics, they dry readily, but the drying temperature must be kept low (<100°C) to prevent damage and shrinkage. Consequently, modacrylics must be ironed below 135°C. Modacrylic fibers are flame retardant and self-extinguishing on removal from a flame, with a LOI of 27. Modacrylic fibers are used in a number of special products including deep pile carpets, fleece and fur fabrics, blankets, scatter rugs, wigs and hair pieces, stuffed toys, and more recently children's sleepwear.

OTHER ACRYLICS

Nytril

Nytril fibers are made up of polymers containing at least 85% vinylidene dinitrile units, which appear at least every other unit in the polymer chain. The comonomer used in Nytril synthesis is vinyl acetate. The two monomers are polymerized in benzene using peroxide catalyst. The polymer is precipitated, washed, and dissolved in N,N-dimethylformamide and then passed through a spinneret into an aqueous coagulating bath to form Nytril fibers. The properties of the Nytril fiber are similar to the other acrylic fibers. The fiber possesses moderate tensile, regain, and thermal properties. The fiber is chemical and sunlight resistant but is as flammable as cellulosic fibers. Darvan Nytril fibers were produced in the U.S. until 1961.

Lastrile

Lastrile fibers are fibers formed from copolymers of acrylonitrile and a diene such as butadiene and contain 10%-50% acrylonitrile units. Lastrile fibers have not been commercially produced. The major proportion of butadiene within the copolymer suggests that the fiber would have extensive elastomeric character.

9. Polyolefin Fibers

Polyolefin fibers are those fibers produced from polymers formed by chain growth polymerization of olefins (alkenes) and which contain greater than 85% polymerized ethylene, propylene, or other olefin units. In general, linear high-density stereoregular polyethylene and polypropylene are used in textile applications, with polypropylene predominating due to its superior temperature stability. These fibers have good strength and toughness, have good abrasion resistance, and are inexpensive. The fibers are difficult to dye and have relatively low melting points, but they are effectively used in a wide variety of textile applications.

Structural Properties

Linear polyethylene and polypropylene are polymerized from their corresponding monomers to a DP of 1000-3000 using complex metal (Ziegler-Natta) catalysts at 40°-100°C and at moderate pressures. Using this initiation technique, branching of the polyolefin due to free radical chain transfer is avoided and a linear unbranched structure is formed. With polypropylene, catalyst systems are selected that will lead to regular isotactic placement (>90%) of the optically active methyl-substituted carbon on the backbone to give a structure capable of high crystallinity:

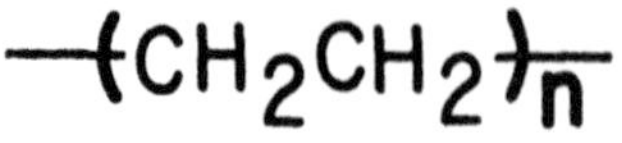

$$-(CH_2CH_2)_n$$

POLYETHYLENE

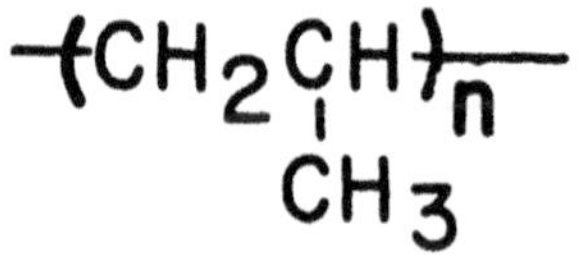

$$-(CH_2CH)_n-$$
$$\qquad\quad CH_3$$

POLYPROPYLENE

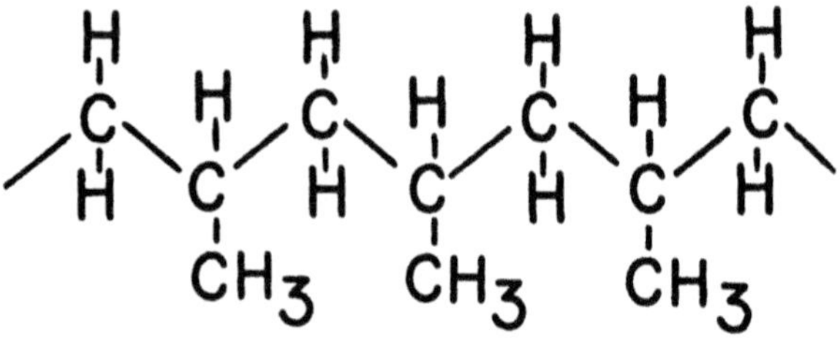

ISOTACTIC POLYPROPYLENE

The resultant polymers are dried, compounded with appropriate additives, melt spun into fibers, and drawn to orient. The highly linear chains of these polyolefins can closely pack and associate with adjacent chains through van der Waals interactions and possess crystallinities in the 45%-60% range. The surface of these fibers is usually smooth, and the fiber cross section is round (Figure 9-1). Also, polyolefin films can be split (fibrillated) using knife edges to form flat ribbonlike fibers with a rectangular cross section.

Physical Properties

Polypropylene and polyethylene are strong fibers with good elongation and recovery properties. The tenacity of fibers varies from 3.5 to 8 g/d (31-81 g/tex), with an elongation at break of 0%-30%. The fibers recover well from stress, with 95% recovery at 10% elongation. The fibers are moderately stiff and have moderate resiliency on bending. Moisture does not affect these properties, since polyolefins are hydrophobic and have a moisture regain of 0%. The polyethylene fibers have specific gravities of 0.95-0.96 and polypropylene specific gravities of 0.90-0.91. As a consequence these fibers float on water and are the lightest of the major fibers in commerce.

Figure 9-1. Polypropylene. x1000.

The fibers are unaffected by solvents at room temperature and are swollen by aromatic and chlorinated hydrocarbons only at elevated temperatures. They exhibit excellent heat and electric insulation characteristics and are extensively used in these applications. The fibers are heat sensitive. Polyethylene softens at 130°C and melts at 150°C, while polypropylene softens at about 150°C and melts at about 170°C.

Chemical Properties

The polyolefins are extremely inert and resistant to chemical attack. They are unaffected by chemical and biological agents under normal conditions. They are sensitive to oxidative attack in the presence of sunlight due to formation of chromophoric keto groups along the hydrocarbon chain. These groups act as photosensitizers for further decomposition. The fiber only slowly undergoes oxidative decomposition at its melting point.

End-Use Properties

Common trade names for polyolefin fibers include Herculon, Marvess, and Vectra. Polyolefins are lustrous white translucent fibers with good draping qualities and a characteristic slightly waxy hand. They have excellent abrasion resistance and exhibit fair wrinkle resistance. They are essentially nonabsorbent, but reportedly exhibit wicking action with water. The polyolefins are strong, tough fibers which are chemically inert and resistant to oxidative attack except that induced by sunlight. Stabilizing chemicals must be incorporated into the fibers to lower their susceptibility to such light-induced attack. The polyolefins have high affinity for oil-borne stains, which are difficult to remove on laundering. The fibers are dry-cleanable with normal dry-cleaning solvents if the temperature is kept below 50°C. Since the fibers have no affinity for water, they dry readily, but care must be taken to keep the drying temperature below 70°C for polyethylene and 120°C for polypropylene to prevent heat-induced thermoplastic shrinkage and deformation. The fibers are flammable, burn with a black sooty flame, and tend to melt and draw away from the flame.

The polyolefins--particularly polypropylene--have found a number of applications particularly in home furnishings and industrial fabrics. Uses include indoor-outdoor carpeting, carpet backing, upholstery fabrics, seat covers, webbing for chairs, nonwovens, laundry bags, hosiery and knitwear (particularly as a blended fiber), fishnet, rope, filters, and industrial fabrics.

10. Vinyl Fibers

Vinyl fibers are those man-made fibers spun from polymers or copolymers of substituted vinyl monomers and include vinyon, vinal, vinyon-vinal matrix (polychlal), saran, and polytetrafluoroethylene fibers. Acrylic, modacrylic and polyolefin--considered in Chapters 8 and 9--are also formed from vinyl monomers, but because of their wide usage and particular properties they are usually considered as separate classes of fibers. The vinyl fibers are generally specialty fibers due to their unique properties and uses. All of these fibers have a polyethylene hydrocarbon backbone with substituted functional groups that determine the basic physical and chemical properties of the fiber.

VINYON

Vinyon is defined as a fiber in which at least 85% of the polymerized monomer units are vinyl chloride. Vinyon fibers have high chemical and water resistance, do not burn, but do melt at relatively low temperatures and dissolve readily in many organic solvents, thereby limiting their application.

Structural Properties

Vinyon is formed by emulsion polymerization of vinyl chloride or by copolymerization of vinyl chloride with less than 15% vinyl acetate in the presence of free radical catalysts to a DP of 200-450:

$$\left[(CH_2CH)_{\overline{x}}(CH_2CH)_{\overline{y}} \right]_n$$

$$\begin{array}{ccc} & & \\ Cl & & O \\ & & \| \\ & & OCCH_3 \end{array}$$

$$x > 85\%, \quad y < 15\%$$

VINYON

The polymer is precipitated and isolated by spray drying and then dissolved in acetone; the polymer solution is dry spun to form the fiber. Comonomers such as vinyl acetate are added to reduce the crystallinity of the drawn fiber and to increase the amorphous areas within the fiber. Weak hydrogen bonding between chlorine and hydrogen on adjacent vinyon chains would be expected with tight packing of the molecular chains in the absence of co-monomer. The fibers are spun in nearly round or dog-bone cross section (Figure 10-1).

Physical Properties

Vinyon fibers have a strength of 1-3 g/d (9-27 g/tex) in both the wet and dry states; elongations at break vary between 10% and 125%. At low elongations vinyon fibers recover completely from deformation. Vinyon fibers are soft and exhibit good recovery from bending deformation. The fiber has moderate density, with a specific gravity of 1.33-1.40. Vinyon is extremely hydrophobic, having a moisture regain of 0.0%-0.1% under standard conditions. Vinyon is readily dissolved in ketone and chlorinated hydrocarbon solvents and is swollen by aromatic solvents. The fiber is a poor heat and electrical conductor and possesses potential in insulation applications. Unfortunately, the oriented fiber softens and shrinks at temperatures above 60°C, thereby limiting the applications in which it may be used.

Figure 10-1. Vinyon. x1400.

Chemical Properties

Vinyon fiber is chemically inert and possesses chemical properties similar to polyolefin fiber. Vinyon is only slowly attacked by ultraviolet rays in sunlight. Vinyon fiber melts with decomposition at 135°-180°C, with vinyon containing comonomer having a lower melting/decomposition temperature.

End-Use Properties

Vinyon as pure polyvinyl is marketed as PVC-Rhovyl, while vinyon HH is a copolymer. The fiber is of low strength but has properties that make it useful in apparel where heat is not a factor. It is difficult to dye. It may be laundered readily, but it is attacked by common dry-cleaning solvents. Since the fiber may not be heated above 60°C, it may be tumble dried only at the lowest heat settings. The fiber is nonflammable and does not support flaming combustion (LOI of 37). Vinyon finds its major use in industrial fabrics including filters, tarps, and awnings, in protective clothing, and in upholstery for outdoor furnishings. It is also used as a bonding fiber in heat bonded nonwovens and as tire cord in specialty tires.

VINAL

Vinal fibers are made from polymers containing at least 50% vinyl alcohol units and in which at least 85% of the units are combined vinyl alcohol and acetal crosslink units. The fiber is inexpensive, resembles cotton in properties, and is produced in Japan.

Structural Properties

Vinal fibers are formed from inexpensive starting materials through a complex process. Vinyl acetate monomer is solution polymerized to polyvinyl acetate using free radical initiation techniques. The polyvinyl acetate is hydrolyzed to polyvinyl alcohol in methanol under basic conditions. Aqueous polyvinyl alcohol solution is wet or dry spun followed by exposure to an aldehyde such as formaldehyde or benzaldehyde and heat (240°C) to form acetal crosslinks between chains to insolubilize the fiber. The polyvinyl alcohol chains are tightly packed except near the juncture of the acetal crosslinks. Extensive hydrogen bonding between hydroxyl groups on adjacent chains occurs, which contributes to its highly packed structure, and the fiber is about 50% crystalline:

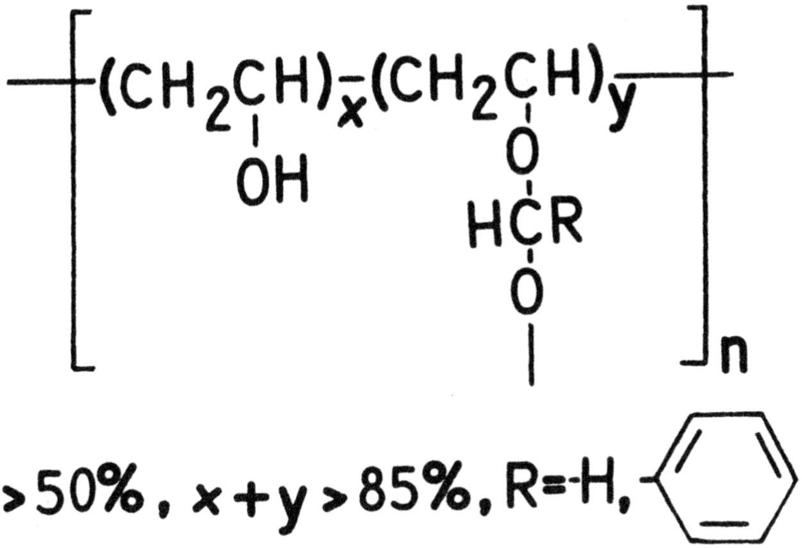

$$\left[(CH_2CH)_{\overline{x}}(CH_2CH)_{\overline{y}} \right]_n$$

OH

O

HCR

O

$x > 50\%$, $x + y > 85\%$, $R = H$, ⬡

VINAL

The fiber surface is somewhat rough with lengthwise striations and possible periodic twists. The fiber may be nearly round or U-shaped in cross section with a noticeable skin and core.

Physical Properties

Vinal fiber is moderately strong with a dry tenacity of 3-8.5 g/d (27-77 g/tex) and possesses a moderate extension at break of 10%-30%. On wetting the strength of the fiber decreases and the elongation at break increases slightly. The fiber exhibits poor to moderate (75%-95%) recovery from elongations as low as 2%. The fiber is moderately stiff, but does not recover readily from dry or wet deformation. The fiber has a specific gravity of 1.26-1.30. Vinal fiber has a moisture regain of 3.0%-5.0% under standard conditions, and is swollen and attacked by aqueous phenol and formic acid solutions. The fiber is a moderate heat and electrical conductor. On heating to 220°-230° the fiber undergoes 10% or more heat-induced shrinkage.

Chemical Properties

Vinal is somewhat sensitive to acids and alkalies. It is attacked by hot dilute acids or concentrated cold acid solutions with fiber shrinkage and is yellowed by strong aqueous solutions of alkali. Oxidizing, reducing, and biological agents have little effect on vinal. Sunlight causes vinal to slowly lose its strength with perceptible changes in color. At 230°-250° vinal shrinks and softens with decomposition.

End-Use Properties

Vinal resembles cotton and other cellulosics in end-use properties. Kuralon and Manryo are names under which vinal fibers are marketed. Vinal fiber has good strength and excellent abrasion and pilling resistance. Like cellulosics, vinal breaks at low elongations, exhibits poor recovery from small deformations, and wrinkles readily unless treated with durable press resins of the type used for cellulosics. Fabrics of vinal have a warm comfortable hand, are absorbent, and exhibit good drapability. The fiber has a silklike appearance and luster. It has excellent sunlight resistance and fair heat resistance. It dyes readily with dyes for cellulosics. Its launderability and dry-cleanability are very good, and it dries more readily than cellulosics. Vinal may be ironed dry up to 120°C, but undergoes severe deformation if ironed wet below 120°C. Vinal burns readily like a cellulosic fiber and has a LOI of 20. Vinal is used in apparel as well as industrial applications where an inexpensive fiber of moderate strength and properties is desired. It is useful in fiber blends to increase comfort and aesthetic properties.

VINYON-VINAL MATRIX FIBER

In response to the need for a fiber of low flammability (LOI of 31) and low toxic gas formation on burning, Kohjin company developed and marketed a vinyon-vinyl (50:50) matrix (polychlal) fiber under the trade names Cordela and Cordelan. The fiber is believed to be formed through grafting of vinyl chloride to polyvinyl alcohol followed by mixing of the resultant copolymer with additional polyvinyl alcohol. The polymer mixture is wet spun, oriented, and crosslinked with aldehydes. The fiber has a kidney-shaped cross section, and no outer skin is evident. The fiber is moderately strong with a dry tenacity of about 3 g/d (27 g/tex) and wet tenacity of 2.2 g/d (20 g/tex). The elongation at break is 15%-20% wet or dry, with poor recovery from even low elongations. It has a specific gravity of 1.32 and moisture regain of 3.0%. Besides its low flammability, it has good chemical resistance except under extreme conditions. The fiber is heat sensitive above 100°C, but it may be dyed at lower temperatures with a large range of dyes for cellulosics as well as man-made fibers. Vinyon-vinyl matrix fiber is used primarily in applications where low flammability in apparel is desired and has been used extensively in children's sleepwear.

SARAN

Saran is the generic name for fibers made from synthetic copolymers that are greater than 80% vinylidene chloride. The fiber is formed through emulsion copolymerization of vinylidene chloride with lesser amounts of vinyl chloride using a free radical catalyst, and the precipitated copolymer is melt spun into the fiber:

$$\left[(CH_2\underset{\underset{Cl}{|}}{\overset{\overset{Cl}{|}}{CH}})_x (CH_2\underset{\underset{Cl}{|}}{CH})_y \right]_n$$

$$x > 80\% , \quad y < 20\%$$

$$SARAN$$

The fiber is of moderate crystallinity and resembles vinyon in properties. The fiber has wet and dry strength of 1-3 g/d (9-27 g/tex), elongation at break of 15%-30%, and good recovery from less than 10% elongation. The fiber has a specific gravity of 1.7 and has excellent resiliency. It has essentially no affinity for water, but it is soluble or attacked by cyclic ethers, ketones, and aromatic solvents. It is a good heat and electrical insulator, but it softens at 115°-160°C and melts at 170°C. The fiber is chemically inert and only slowly affected by the ultraviolet rays in sunlight. The fiber is of low flammability. Saran is marketed as saran or under the trade name Rovana. The fiber has high resiliency and low water absorbency and is chemically inert and used in applications where these qualities are desirable and where heat sensitivity is not a problem. The fiber is generally not used in apparel but rather in automobile upholstery, outdoor fabrics, home furnishings, and industrial applications.

POLYTETRAFLUOROETHYLENE

Tetrafluoroethylene is better known by its trade name--Teflon--and is widely used in many applications including specialty fibers. Polytetrafluoroethylene fiber is extremely hydrophobic and chemically and thermally stable and is used in applications where such stability and inertness is needed. Polytetrafluoroethylene fibers are produced through polymerization of tetrafluoroethylene under conditions of high temperature and pressure using peroxide catalysts. The polymer is spun as a dispersion to form a weak fiber which is then heated at 385°C to fuse the individual fiber particles. The fiber must be bleached to give a white fiber. The highly electronegative fluorine atoms in the fiber result in an extremely tight packing of adjacent molecular chains tightly held by van der Waals forces and with a high degree of crystallinity. The fibers are smooth with a round cross section.

The fiber has a tenacity of 1-2 g/d (9-18 g/tex) and elongations at break vary from 15% to 30%. It is a soft flexible fiber of high density (specific gravity 2.1), and it is totally hydrophobic. It is unaffected by solvents except for perfluorinated hydrocarbons above 300°C. It is an excellent electrical and heat insulator and is not affected by heat up to 300°C. It is chemically inert and may be used in numerous industrial end-use applications including protective clothing. Its hydrophobicity has permitted is to be used to form breathable but water repellant composite materials for textile usage, particularly in outdoor and rainwear.

$$-(CF_2CF_2)_n-$$

TETRAFLUOROETHYLENE

11. Elastomeric Fibers

Elastomeric fibers are those fibers that possess extremely high elongations (400%-800%) at break and that recover fully and rapidly from high elongations up to their breaking point. Elastomeric fibers are made up of molecular chain networks that contain highly amorphous areas joined together by crosslinks. On elongation, these amorphous areas become more oriented and more crystalline in nature. Elongation continues until the crosslinks in the structure limit further molecular movement. If additional force is placed on the elastomeric fibers at this point, molecular scission occurs, causing a reduction in properties and ultimate breaking of the fiber. Elastomeric fibers include the crosslinked natural and synthetic rubbers, spandex fibers (segmented polyurethanes), anidex fibers (crosslinked polyacrylates) and the side-by-side biconstituent fiber of nylon and spandex (Monvelle). The fibers are all used in specialized applications where high elasticity is necessary within the textile structure.

RUBBER

Rubber fibers from natural sources have been known for over 100 years. Natural rubber in commerce is derived from coagulation of Hevea brasiliensis latex and is primarily cis-polyisoprene, a diene polymer. Most synthetic rubbers were developed during and following World War II. They are crosslinked diene polymers, copolymers containing dienes, or amorphous polyolefins. Both the natural and synthetic rubbers must be crosslinked (vulcanized) with sulfur or other agents before true elastomeric properties are introduced. In addition, accelerators, antioxidants, fillers, and other materials are added to the polymeric rubber prior to fiber formation.

Rubber fibers exhibit excellent elastic properties but are sensitive to chemical attack, thereby limiting their usefulness.

Structural Properties

Rubber fibers are derived from the sources outlined above and structurally are crosslinked polyisoprene, polybutadiene, diene-monomer copolymers, or amorphous polyolefins. Less common or special use rubbers include the acrylonitrile-diene copolymer Lastrile and the chloroprene polymer Neoprene. Structural formuli of typical rubbers follow:

$$\left[\begin{array}{c} CH_2 \\ | \\ CH_3 \end{array} C=C \begin{array}{c} CH_2 \\ | \\ H \end{array} \right]_n \begin{array}{c} CH- \\ | \\ S_{1-10} \\ | \end{array} \qquad \overline{\left(CH_2 - CH=CH-CH_2 \right)_n}$$

**CROSSLINKED cis–POLYISOPRENE
(VULCANIZED NATURAL RUBBER)**   **POLYBUTADIENE
(BUTYL RUBBER)**

$$\left[\begin{array}{c} CH_2 \\ \\ H \end{array} C=C \begin{array}{c} CH_2-CH_2 CH- \\ \\ H \end{array} \right]_n$$

**cis–POLYBUTADIENE–STYRENE COPOLYMER
RUBBER**

The rubber polymers emulsified as a latex are blended with vulcanizing agents, chemical accelerators, antioxidants, fillers, and other materials. The polymer blend is extruded or coagulated and cut into fibers and heated to crosslink the structure. Of necessity, the rubber fibers are large in cross section and have a higher linear density compared to other man-made fibers. The hydrocarbon chains in the crosslinked rubbers are in a highly folded but random and amorphous configuration, but they are attached periodically to adjacent polymer chains through multiple sulfur or other crosslinks. On stretching, the molecular chains in the amorphous region untangle and straighten and orient into parallel, more crystalline struc- tures, up to the elastic limit determined by the crosslinks present. On relaxation, the molecular chains return to their lower-energy amorphous state. Rubber fibers tend to possess a square or round cross section de- pending on whether the fibers were extruded or cut.

Physical Properties

Rubber fibers possess tenacities of only 0.5-1 g/d (4.5-9 g/tex) wet or dry. Elongation at break for these elastomeric fibers varies between 700% and 900%, with nearly complete elastic recoveries even at higher elongations. Because of their size, rubber fibers are moderately stiff but resilient. They are light, having a specific gravity of only 0.95-1.1.

Certain environmental properties of rubber limit its usefulness. Be- ing hydrocarbon material, rubber fibers do not absorb moisture and have a regain of 0.0% under standard conditions. The fiber is a poor heat conduc- tor and a good heat and electrical insulator. The fiber is readily swollen by ketones, alcohol, hydrocarbons, and oils and softens if heated above 100°C.

Chemical Properties

Rubber fibers are quite susceptible to attack by chemicals both in solutions and in the environment due to the presence of extensive unsatur- ation in the molecule. Rubber is readily attacked by concentrated sulfuric and nitric acids and by oxidizing agents such as hydrogen peroxide and sodium hypochlorite at elevated temperatures. Although biological agents don't readily attack rubber, heating above 100°C or exposure to sunlight accelerates its oxygen-induced decomposition. Rubber fibers are particu- larly susceptible to attack by atmospheric ozone (O_3), a major component of smog.

End-Use Properties

Rubber fibers are manufactured by several companies as rubber fiber or under trade names such as Buthane, Contro, Hi-Flex, Lactron, Lastex, and Laton. Rubber fibers show good elastomeric properties and reasonable aes- thetic properties particularly as the core of a textile yarn structure. Rubber fibers usually have a dull luster due to the additives and fillers within the fiber. The fiber has poor resistance to household chemicals, sunlight, heat, and atmospheric contaminants. It cannot be dyed readily, and the polymer must be colored prior to fiber extrusion and curing. The fiber possesses only fair launderability and poor dry-cleanability. Owing to its low moisture absorbance, it dries readily, but care must be taken to dry or iron the fiber below 100°C. Rubber fibers burn readily but tend to be self-extinguishing. Rubber fibers are used extensively in applications where elastomeric fibers are desired, such as in swimwear, foundation garments, outerwear, and underwear, hosiery, and surgical bandages. The use of rubber fibers has decreased in recent years as more stable elastomeric fibers such as spandex have been introduced.

SPANDEX

Spandex fibers are elastomeric fibers that are >85% segmented polyurethane formed through reaction of a diisocyanate with polyethers or polyesters and subsequent crosslinking of polyurethane units. The spandex fibers resemble rubber in both stretch and recovery properties, but are far superior to rubber in their resistance to sunlight, heat, abrasion, oxidation, oils, and chemicals. They find the widest use of any of the elastomeric fibers.

Structural Properties

Spandex is a complex segmented block polymer requiring a complex series of reactions for formation. Initially, low molecular weight polyethers and polyesters (oligomers) containing reactive terminal hydroxyl and/or carboxyl groups are reacted with diisocyanates by step growth polymerization to form a capped prepolymer. This polymer is melt spun or solvent spun from N,N-dimethylformamide into a fiber; then the fiber is passed through a cosolvent containing a reactive solvent such as water that reacts with the terminal isocyanite groups to form urethane crosslinks. Typical structures for spandex with urethane crosslinks follow:

$$\left(\mathrm{NHCNH-R'-NHCO-R-CNH-R'}\right)_n$$

$$R = \left(\mathrm{CH_2CH_2O}\right)_n \, , \, \left(\mathrm{CH_2}\right)_6 \mathrm{CO} \, , \text{ etc.}$$

R' = ⬡ , ⬡-CH₂-⬡ , ⬡-CH₃ , etc.

SPANDEX

The polyether or polyester segments in spandex are amorphous and in a state of random disorder, while urethane groups segmenting the polyether or polyester segments can form hydrogen bonds and undergo van der Waals interactions with urethane groups on adjacent chains. Chain ends will be cross-linked or joined to other chains through urea groups. On stretching, the amorphous segments of the molecular chains become more ordered up to the limit set by the urea linkages. On relaxation, the fiber returns to its original, less ordered state. Spandex fibers are long, opaque fibers available in many cross sections including round, lobed, or irregular (Figure 11-1).

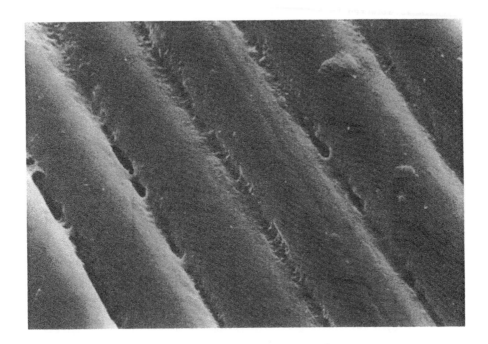

Figure 11-1. Spandex. x1200.

Physical Properties

Spandex fibers are weak but highly extensible fibers with tenacities of 0.5-2 g/d (4.5-18 g/tex) dry with slightly weaker wet tenacities, and elongations at break of 400%-700%. Spandex exhibits nearly complete recovery from high elongations and is quite resilient. The fibers have specific gravities of 1.2-1.4 and are somewhat more dense than rubber fibers. Owing to the more hydrophilic nature of spandex, its moisture regain varies between 0.3% and 1.3% under standard conditions. Although spandex is swollen slowly by aromatic solvents, it is generally unaffected by other solvents. Spandex is a heat insulating material and shows poor heat conductivity. It softens between 150°C and 200°C and melts between 230°C and 290°C. It has moderate electrical resistivity and builds up some static charge under dry conditions.

Chemical Properties

Spandex fibers are much more stable to chemical attack than rubber. Spandex is attacked by acids and bases only at higher concentrations and under more extreme reaction conditions. Reducing agents do not attack spandex, but chlorine bleaches such as sodium hypochlorite slowly attack and yellow the fiber. Spandex generally resists attack by organic solvents and oils and is unaffected by biological agents. Sunlight causes slow yellowing and deterioration of spandex, and heat causes slow deterioration at elevated temperatures.

End-Use Properties

Spandex is widely used and sold as an elastomeric fiber under that designation or under such trade names as Cleerspan, Glospan, Lycra, or Numa. Spandex possesses excellent elastomeric properties and acceptable aesthetics for use in constructions requiring such a fiber. Spandex is dull, but luster may be improved with brighteners. Although it has low water absorbency, spandex has good resistance to abrasion and wrinkling and may be heat set if desired. It has good resistance to household chemicals other than bleach and is resistant to sunlight or heat-induced oxidation. It can be dyed relatively easily to reasonable degrees of colorfastness. It stands up well to repeated laundering and dry-cleaning, and it dries readily without damage at moderate temperatures (less than 120°C). Spandex burns readily but melts and shrinks away from the flame. Spandex can be used effectively in all elastomeric fiber applications.

OTHER ELASTOMERIC FIBERS

Anidex

In 1970, anidex fibers were introduced as an elastomeric fiber by Rohm and Haas with the trade name Anim. Anidex fibers are defined as fibers containing polymers that are at least 50% of one or more polymerized acrylate esters. Anidex fibers are formed through emulsion copolymerization of acrylate esters with reactive crosslinkable comonomers such as N-methylolacrylamide. The resulting copolymer emulsion is mixed with a filler and wet spun to form a fiber which is heated to crosslink the polymer chains and provide the necessary elastomeric properties. The morphology and elastomeric action of the fiber resemble spandex and rubber, but anidex generally has a lower elongation at break than these fibers. It has a round cross section. The fiber has a specific gravity of 1.22 and a moisture regain of 0.5% under standard conditions. Anidex is reported to be more resistant to heat, light, and chemicals than either spandex or rubber. Otherwise, the fiber possesses end-use properties very much like those of spandex. Anidex fibers apparently did not have sufficient differences in properties to become an economic success and are no longer being produced.

Nylon-Spandex Biconstituent Fiber

An inherent problem with elastomeric fibers is their low strength and limited abrasion resistance. A new biconstituent elastomeric fiber-- Monvelle--has been introduced to answer these problems. The fiber is a side-by-side bicomponent fiber containing nylon and spandex. The fiber is melt spun by special spinnerets to form the fiber. On drawing, the nylon portion of the fiber stretches and becomes more molecularly oriented, whereas the spandex simply elongates as would be expected for an elastomeric fiber. On relaxation, the spandex portion contracts, whereas the nylon portion does not, and the fiber coils into a tight coil which will have excellent elastomeric properties and high strength even as the spandex approaches its elastic limit. Sheer fibers may be formed from this bicomponent fiber, and the fiber is readily dyeable due to its nylon content. The fiber is particularly useful in support hose and in pantyhose and other constructions where elasticity and strength are important.

12. Mineral and Metallic Fibers

A number of fibers exist that are derived from natural mineral sources or are manufactured from inorganic and mineral salts. These fibers are predominantly derivatives of silica (SiO_2) or other metal oxides. In addition, metal fibers (either alone or encapsulated in a suitable organic polymer) are produced. The common feature of these fibers is their inorganic or metallic composition and tendency to be heat resistant and non-flammable, with the exception of polymer-coated metallic fibers.

GLASS

Fibers spun from glass are completely inorganic in nature and possess unique properties that cannot be found in organic textile fibers. Glass fibers have some deficiencies in properties that severely limit their use in apparel. Glass fibers are used in a number of industrial and aerospace applications and in selected home furnishing uses where heat and environmental stability are of prime importance.

Structural Properties

Glass fibers are formed from complex mixtures of silicates and borosilicates in the form of mixed sodium, potassium, calcium, magnesium, aluminum, and other salts. The silicate-borosilicate prepolymer is prepared through mixing and fusing the following inorganic salts in the range of concentrations indicated:

Silica	50%-65%
Calcium oxide	15%-25%
Alumina	2%-18%
Boron oxide	2%-15%
Other oxides	1%-10%

The exact composition of the manufactured glass polymer will affect the ultimate properties of the glass fiber formed. The glass thus formed as small marbles is melted in an electric furnace at high temperatures (>800°C) and melt spun to form fine smooth glass fiber filaments (Figure 12-1). The resultant silicate-borosilicate chains within the fiber are not ordered, and the glass structure is totally amorphous. Glass is actually a supercooled liquid that exhibits extremely slow but observable flow characteristics with time. The polymer chains within a glass fiber can be represented as follows:

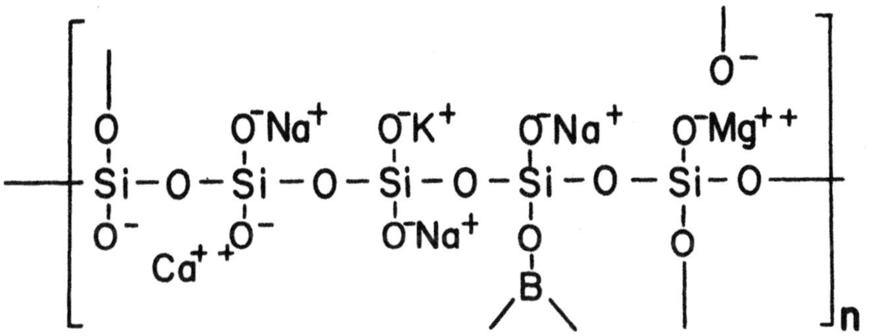

GLASS

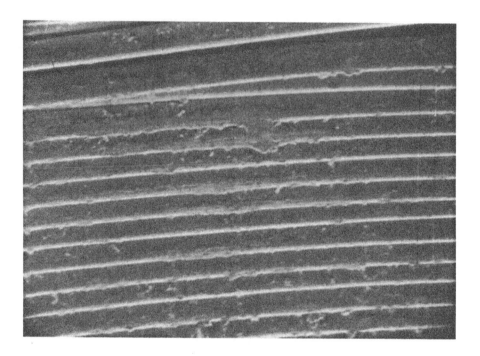

Figure 12-1. Glass. x1100.

Physical Properties

Glass fibers are extremely strong, with tenacities of 6-10 g/d (54-90 g/tex) dry and 5-8 g/d (36-72 g/tex) wet. They possess elongations at break of only 3% or 4% but are perfectly elastic within this narrow deformation range up to their breaking point. Glass fibers are quite stiff and brittle and break readily on bending. As a result, they exhibit poor resistance to abrasion, although appropriate organic sizings can alleviate this problem to some extent. Glass fibers are extremely dense, having a specific gravity of about 2.5.

The surface of glass fibers can be wetted by water, but otherwise the fibers have essentially no affinity for water. As a result the moisture regain of these fibers is 0.5% or less. Glass fibers are not soluble in common organic solvents but can be slowly dissolved by concentrated aqueous base and more rapidly by hydrofluoric acid. Glass fibers exhibit excellent heat and electrical insulation properties and are not affected by heat up to their melting point of 750°C or above.

Chemical Properties

Glass fibers are chemically inert under all but the most extreme conditions. Glass fibers are attacked and etched by dilute hydrofluoric acid and by concentrated alkalies over long periods of exposure. Oxidizing and reducing agents and biological agents have no effect on glass fibers under normal conditions, and glass fibers are unaffected by sunlight or heat.

End-Use Properties

Glass fibers are manufactured for industrial and consumer use under a number of names including fiberglas, Beta glass, J-M fiberglass, PPG fiberglass, and Vitron. Glass fibers are strong, but they exhibit poor abrasion resistance, which limits their use in textile structures for consumer goods. Heat setting of glass fabrics imparts good wrinkle resistance to fabrics made from these fibers. Glass fabrics have low moisture absorbency and good hand and luster when used in home furnishing applications. The high density of glass provides reasonable draping properties for most applications, although glass fabric constructions are extremely heavy in weight. Its sunlight and heat resistance make glass fibers useful in window coverings. Glass fibers are extremely difficult to dye, and pigment-binder or pigment-melt coloration must be used. Glass substrate must be laundered or dry-cleaned using minimum mechanical action and drip drying. Glass substrates do not need ironing, although careful ironing up to 400°C is possible. Glass and related inorganic fibers are the only truly non-flammable man-made fibers. They are used extensively in curtains and draperies, electrical and thermal insulation, tire cord, reinforced plastics, industrial filters, and protective clothing and accessories. Apparel applications are limited to very fine Beta glass fibers.

INORGANIC FIBERS

A series of man-made inorganic fibers other than glass exist that are nonflammable, heat stable amorphous materials useful in industrial fabric constructions, including refractory materials. These inorganic fibers include pure silica, potassium titanates, aluminum borosilicates, and aluminum oxide-zirconium oxide polymers. Most of these fibers have high strength, are less susceptible to chemical attack than glass, and melt above 1000°C. They may be used in higher-temperature applications than possible with most glass fibers.

ASBESTOS

Asbestos is the name given to several natural minerals (anthophyllite, amphibole, serpentine) which occur in a fibrous crystalline form. The asbestos is initially crushed to open up the fiber mass, followed by carding and spinning to yield fibers of circular cross section 1-30 cm in length. Asbestos is very resistant to heat and burning, to acids and alkalies, and to other chemicals. Although it has low strength, asbestos fiber does not deteriorate in normal usage, and it is not attacked by insects or microorganisms. Asbestos is used in fireproof clothing, conveyor belts, brake linings, gaskets, industrial packings, electrical windings, insulations, and soundproofing materials. Inhaled asbestos fibers have been shown to be a serious health hazard, and it has been removed from the textiles market.

METALLIC FIBERS

Metallic fibers are defined as fibers composed of metal, plastic-coated metal, or metal-coated plastic. Single-component metallic fibers for textile usage are fine drawn filaments of metal which can be spun and woven on normal textile machinery. These metallic fibers possess the properties of the metal from which they are formed. Multicomponent metallic fibers are more commonly used in textiles and are usually made from flat aluminum filaments surrounded with or bonded between clear layers of polyester, cellophane, or cellulose ester or from polyester film which has been metallized through vacuum deposition of aluminum and then encapsulated in polyester. In general, the properties of these fibers resemble the properties of the plastic film used to form the multicomponent fiber. The fibers are generally weak and easily stretched but can be used for decorative purposes and for applications where electrical conductivity and heat resistance are important. Trade names for metallic fibers include Brunsmet and Lurex.

13. Miscellaneous Fibers

In this chapter fibers which do not logically "fit" under other classifications are listed--novaloid, carbon, poly-m-phenylenedibenzimidazole
and polyimide fibers. These fibers were developed for specific industrial
applications and do not find wide use in consumer goods. For the present,
it appears unlikely that new generic fiber types will be introduced in the
marketplace unless they have distinguishing characteristics that make them
uniquely suited to particular applications. All of these fibers possess
such properties.

NOVALOID

Novaloid is the designation assigned by the FTC for a class of flame
retardant fibers made from crosslinked phenol-formaldehyde polymer. The
fibers of this class in U.S. production are called Kynol and are manufactured by American Kynol, Inc. The fiber is golden yellow in color and
possesses good physical and chemical properties. The fiber is thought to
be formed through spinning phenol-formaldehyde prepolymer, followed by
heating the fiber in formaldehyde vapor to crosslink the structure. The
fiber has low crystallinity with a tenacity of 1.5-2.5 g/d (14-23 g/tex)
and an elongation at break of about 35%. The fiber has a specific gravity
of 1.25 and a moisture regain of 4%-8%. It is inert to acids and organic
solvents but is more susceptible to attack by bases. The fiber is totally
heat resistant up to 150°C. It is inherently flame retardant and chars
without melting on exposure to a flame. LOI of the fiber is very high
(36). This characteristic yellow fiber is dyeable with disperse and
cationic dyes. Novaloids are characteristically used in flame retardant

protective clothing and in apparel and home furnishings applications where low fiber flammability is desired.

$$\left[CH_2 - \underset{\underset{O-}{\overset{\overset{O}{|}}{CH_2}}}{\bigcirc} \right]_n$$

NOVALOID

CARBON

Carbon or graphite fibers have been developed recently for use in industrial and aerospace applications. The carbon fibers are prepared from rayon, acrylic, or pitch fibers by controlled oxidation under tension in limited oxygen atmosphere at 300°-400°C. At this stage of oxidation, carbon fibers have sufficient flexibility to be used in apparel applications such as flame retardant clothing. Further oxidation at temperatures near or exceeding 1000°C under tension results in a high strength fiber consisting of a continuous condensed aromatic network similar to graphite. Fully oxidized carbon fibers have tenacities from 10 to 23 g/d (90 to 207 g/tex) dry or wet and elongations at break of just 0.4% to 1.5%. They have specific gravities from 1.77 to 1.96 and are mildly electrically conducting. Although the fibers are of high strength, they exhibit poor abrasion characteristics and must be sized with epoxy or other resins before formulation into textile substrates. Carbon fibers are inert to all known organic solvents and to attack by acids and bases under normal conditions. These black fibers have excellent resistance to sunlight and biological agents, are inherently flame retardant and are resistant to oxidation at high temperatures. Carbon fibers are extensively used in reinforcing fibers for resins and plastics in high performance fiber-polymer composites and are marketed as Celion, Hi-Tex, and Thornel.

POLY-m-PHENYLENEDIBENZIMIDAZOLE (PBI)

PBI was developed by the U.S. Air Force and Celanese as a flame retardant fiber for use in aerospace applications. The fiber is spun from N,N-dimethylacetamide followed by derivatization with sulfuric acid to form a golden fiber. It has moderate strength (3 g/d or 27 g/tex) and good elongation at break (30%), a moderate density (1.4 g/cm^3), and a high moisture regain (15%). Fabrics from the fiber possess good hand and drape and are stable to attack by ultraviolet light. PBI does suffer from poor dyeability, however. In addition to its low flammability, the fiber has a high degree of chemical and oxidation resistance, does not show appreciable heat shrinkage up to 600°C and generates only small quantities of smoke and toxic gases on ignition. In recent years, it has shown potential as a replacement for asbestos, as a flue gas filter material, and as an apparel fabric in specialized applications.

POLYIMIDE

An aromatic polyimide has been introduced by Upjohn Company for use in flame retardant, high-temperature applications. The fiber is spun from the polymer by wet or dry processing techniques using a polar organic solvent such as N,N-dimethylformamide to give a fiber with a round or dog-bone cross section. The highly colored fiber may be crimped at 325°C using fiber relaxation. The fiber has a tenacity of 2-3 g/d (18-27 g/tex) and a 28%-35% elongation at break. The fiber has a moisture regain of 2.0%-3.0% and melts at about 600°C.

III. Yarn and Textile Substrate Formation

14. Yarn Formation

Yarn formation methods were originally developed for spinning of natural fibers including cotton, linen, wool and silk. Since the overall physical characteristics of the fibers and processing factors needed differed from fiber to fiber, separate processing systems were developed. As synthetic fibers were introduced, synthetic spinning systems for texturized and untexturized cut staple were developed as modifications of existing staple systems, whereas spinning systems for texturized and untexturized filament were developed separately. Staple yarn formation involves multiple steps and can include: (1) fiber cleaning and opening (as needed for natural fibers); (2) fiber blending (to assure uniform mixing in natural fibers or in fiber blends); (3) carding (to align fibers and to remove short fibers); (4) combing (if highly aligned fibers are desired); (5) drawing and spinning (to reduced the denier of the yarn, to provide twist and to give cohesion to the yarn); and (6) doubling or plying and twisting of the yarns (as needed to provide greater uniformity). In recent years a number of staple spinning processes other than ring spinning have been developed that reduce or shorten the number of steps necessary for formation of yarns suitable for textile substrate formation and are discussed separately following conventional ring spinning techniques. Yarn preparation from fiber filaments is much less complex and often no or only limited twist is imparted prior to use in the textile substrate. The steps involved in yarn formation are outlined in Figure 14-1.

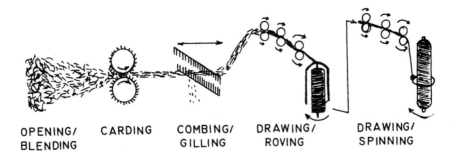

OPENING/   CARDING   COMBING/   DRAWING/   DRAWING/
BLENDING   GILLING   ROVING   SPINNING

Figure 14-1. Spinning.

COTTON SYSTEM

When cotton bales arrive at the textile mill, they are highly compressed and have been ginned to remove seeds and some of the impurities
present. On removing the cotton from the bales, the cotton is opened up,
blended and mixed with cotton from other bales. During the opening and
blending process, the fibers are separated and loosened from each other,
trash is removed from the fibers, and the fibers are more randomly mixed to
assure greater uniformity. Finally the fibers are formed into a thin partially oriented continuous web of intertwined fibers called a picker lap.
The picker lap in turn undergoes carding to remove short fibers and remaining trash and to provide additional orientation to the fibers. Carding
involves pulling, separating, and orienting the fibers by passing the lap
between successive cylinders moving at different speeds and containing fine

bent wire bristles that catch the fibers. The carded lap is removed by a doffer cylinder in the form of sliver (a rope-like fiber mass) and coiled into a rotating can. The sliver may undergo additional blending to improve its uniformity and density so that it may be more effectively drawn and spun into yarn. At this point, the sliver is suitable for drawing and spinning into medium and coarse yarns. Before drawing and spinning into fine yarns, the cotton sliver must first be combed to further straighten and orient the fibers and to remove additional short tangled fibers.

The drawing and spinning process involves passing the sliver between and through a series of rollers moving at progressively higher speeds to draw the sliver to a finer, more oriented, and uniform structure followed by twisting as the sliver is played onto a turning spindle. The drawing portion of the operation is referred to as the drafting process. The degree of twist will depend on the speed of the turning spindle with each complete turn of the spindle providing a single complete twist in the yarn. In the initial stages where little twist is present in the drawn sliver, the sliver is fed through a tube and onto the spindle. This process is called roving. Subsequent drawing and high speed twisting is carried out by ring spinning in which the drafted, lightly twisted sliver (roving) is fed from the drafting unit onto a high speed spindle via a traveller holding the spun yarn to a ring surrounding the reciprocating spindle. The traveller can move easily around the ring and provides a slight drag on the yarn as it is fed onto the spindle. Ring spinning proceeds at 5,000 to 10,000 revolutions per minute.

WOOLEN AND WORSTED SYSTEMS

Wool is spun into yarns by either the woolen or worsted system. The woolen spinning system is less complex than the worsted system and utilizes shorter wools of a wider range of lengths and diameters to give a low twist bulky yarn. The worsted system produces highly twisted fine yarns utilizing fine fibers of a narrow distribution of length and size. Fabrics made from woolen yarns tend to be bulky and contain more entrapped air; whereas fabrics from worsted yarns tend to be tightly woven and fine with a hard flat surface.

On arrival at the mill as bales, raw wool contains large amounts of grease, swint (salts from the body of the sheep), dirt, vegetable material and other impurities. The wool must be washed in successive baths of detergent solution to remove these impurities. The process is called scouring, and the weight of the raw wool can be reduced by as much as 50% by the scouring process. Wool grease (lanolin) is effectively recovered from the scouring liquor as a commercial product. Vegetable matter remaining in the wool can be removed by passing the wool through concentrated sulfuric acid to chemically destroy the cellulosic matter, a process called carborization. After washing and drying, the cleaned wool is blended and carded as described previously to form a sliver. At this point the sliver can be drawn and slightly twisted to form a roving which can then be spun into a woolen yarn. The sliver must undergo additional straightening, orientation, and removal of short fibers to be used in the worsted system. This process involves several successive steps including gilling (a form of pin orientation) and combing to give wool top. The wool top is drawn and slightly twisted in several stages to form a roving which is finally spun into a highly twisted worsted yarn.

OTHER STAPLE SYSTEMS

Other natural staple and synthetic fibers can be spun on cotton and wool systems if these systems are modified to take into account unique factors such as fiber length, crimp and linear density. Cut staple manmade fibers arrive in boxes at the mill and are ready for carding and processing into yarn. When two or more different staple fibers are mixed, it is critical to provide extensive blending before carding and repeated doubling of the sliver to assure intimate blending prior to roving and spinning. Other spinning methods that take fewer steps have been developed for staple spinning and are discussed under other yarn forming methods.

FILAMENT SYSTEMS

Filament spinning systems are much less complex because the fibers are continuous and do not need to be highly twisted to give a cohesive strong yarn. Filament yarn spinning usually involves man-made fibers and only the portion of the ring spinning system that involves twisting and winding onto spindles is used. Other methods are available to give cohesion to a filament yarn and are discussed under other yarn forming systems in the next section.

OTHER YARN-FORMING SYSTEMS

Since conventional ring spinning systems take several steps to reduce fibers to a spun yarn, several alternative techniques have been developed that reduce the number of these steps. Most of these processes have more limited usefulness with regard to the range of linear densities and type and distribution of fiber lengths that can be processed by these systems.

Open-End or Break Spinning

Open-end or break spinning systems have come into wide use for spinning of short and medium staple fibers directly from sliver to yarn in one step without carding or roving. Sliver is fed into a breaking or opening unit to separate the fibers from one another, and the fibers are then forced by air pressure into a hollow rotor rotating at high speeds (up to 50,000 rpm). The fibers are deposited by centrifugal force on the sides of the hollow rotor, and the resulting yarn is removed continuously by a stationary tube mounted within the rotor. The rotating rotor provides twist to the yarn and produces a yarn with somewhat greater higher pitch and bulk and somewhat lower strength than ring spun yarns. Open-end spinning operates at a rate up to five times that of ring spinning and can be effectively used for cotton, polyester-cotton blends, as well as other short and medium staple systems. Synthetic staple fibers such as polyester alone can not be effectively open end spun due to dusting of oligomer from the fibers that interferes with the spinning action of the rotor.

Friction Spinning

Friction spinning is a variation of the open-end spinning system and often referred to as the DREF system. In this system, one or more slivers are fed onto a rapidly rotating card drum which opens the sliver to form single fibers. The separated single fibers are blown from the card drum by a stream of air onto the junction of two parallel perforated drums turning in the same direction. The rotating perforated drums under suction cause the fibers to be compressed and twist around one another to form a uniform yarn which is continuously removed onto a spool. The resultant yarns are bulky and have properties similar to woolen yarns.

Air-Vortex Spinning

Air-vortex spinning is similar to open-end spinning but utilizes a stationary tube rather than a rotor. A high speed air vortex is created in the tube that deposits fibers within the tube and simultaneously provides twist due to the vortex. The yarn is continuously withdrawn as in the case of open end spinning.

Fasciated Spinning

In fasciated spinning long staple sliver without twist is introduced into a limited space and subjected to a torque jet operating at right angles to the flow of the sliver thereby imparting a false twist to the sliver. As the sliver exits the torque jet it rapidly untwists, and the outer fibers tend to break away from the sliver and wrap around the inner sliver to give a strong yarn consisting of mostly parallel fibers with some fibers tightly twisted around the outside.

Self-Twist Spinning

In self-twist spinning, two parallel slivers are fed between two reciprocating rollers which form identical left hand or right hand twists in each of the slivers alternating down the length of the sliver. The resulting false-twisted yarns are then brought together so that the right hand twist segment of one sliver is phased with the left hand twist of the other sliver. On relaxation, the slivers untwist over one another to form a stable yarn. The process was originally developed for spinning wool, but has been used extensively for acrylic yarns. In the related Selfil spinning method the self-twist yarn is wrapped by alternating phases of continuous filaments to form a highly stable wrapped yarn containing less than 10% filament.

Coverspun Spinning

Coverspun is both the name of a spinning system and a yarn. In this system staple rovings are drafted in a conventional manner, and then the roving is passed into a hollow vertical spindle. On the outside of the spindle a filament yarn, from a cylindrical spool rotating at 20,000 to 30,000 rpm, is fed into the top of the hollow vertical spindle with the roving. The rotating filament spindle causes the filament to wrap around the core of the staple to produce a wrapped yarn consisting of 80% to 95% staple. Polyester filaments are usually used to wrap pure yarns or blends of cotton, wool, nylon, and acrylic staple to form sewing threads or yarns for textile substrate production.

Integrated Composite Yarn Spinning

A composite yarn is formed by melt extruding fibers from a spinneret or by coating filaments with a molten polymer followed by coating the emerging fibers with short staple fibers. The resulting matrix is immediately twisted causing the staple fibers to imbed in the extruded fibers before cooling. The resulting composite yarn is formed at a very rapid rate on the Bobtex spinning apparatus developed for this process.

Twistless Systems

In twistless systems liquid or powdered polymer adhesives are applied to sliver or filament tow and the adhesive activated by heating or steam to cause the individual fibers to adhere to one another. In some systems, after textile substrate formation from the twistless yarns the adhesive is removed to improve the aesthetics of the resulting textile. Adhesives used include polyvinyl acetate, polyvinyl alcohol, and starch.

15. Textile Substrate Formation

Yarns often must undergo additional processing before they are ready for use in forming of a textile substrate. The yarns may need to be re-wound onto appropriate packages, reinforced by application of size, lubri-cated by application of spinning oil, and/or drawn-in and tied into the machine used in fabric forming. The combined process used for size appli-cation and lubrication of warp yarns in weaving and warp-knitting is called slashing.

Winding

Winding processes involve movement of yarn from one package to another and often conversion of the overall size, shape and tightness of the pack-ages. These processes also serve other important functions. Winding allows clearing of the yarn to eliminate thin spots, thick spots, knots, and other imperfections, and makes it possible to regulate tension within the package, combine or segment yarn packages, and prepare packages for dyeing prior to substrate formation. In shuttle weaving, it is necessary to prepare small packages referred to as quills or pirns that fit within the shuttle. The yarn is wound onto the pirn sequentially in such a way to assure steady and even release of yarn from the pirn during the weaving process.

Warping and Slashing

A specialized type of package formation is involved in preparing warp beams for weaving or warp knitting. A high degree of tension is placed on the warp during these processes, therefore the yarns must be lubricated to minimize friction between yarn and machine parts and adhesive must be applied to the yarn to strengthen and reduce the hairiness of the yarn. Warping involves winding yarns from several thousand packages placed on creels onto a flanged beam passing through a reed (a comb-like device). The reed maintains the yarns parallel to one another as they are wrapped onto the beam under as even a tension as possible. Warping of small sections of warp (tape warping) is also often carried out, and the tape warps are later placed parallel to one another to provide a full width warp for use in the loom or warp knitting machine.

Staple yarns and some filament yarns must undergo slashing. Slashing involves simultaneous application of sizing and lubricant to the warp from one bath called a size box, followed by drying to remove water or solvent, breaking the slashed warp yarns away from one another using least rods, and rewinding of the warp. Sizes and lubricants used on warps will vary with fiber type. Sizes used include starches and gums, cellulose derivatives such as carboxymethyl cellulose, proteins, polyvinyl alcohol, polyvinyl acetate, and acrylic copolymers, while the lubricants used are similar to spinning oils and include mineral and vegetable oils and waxes as well as derivatives of these materials.

Drawing-In and Tying-In

After warp beams are prepared, the warp yarns must be drawn through certain elements in the loom or warp knitting machine before fabric can be produced. This process in the past was carried out by hand using a special hooked wire to draw each yarn through the elements of the loom or wrap knitter followed by hand-knotting of the yarn to the corresponding yarn on the take-up warp beam. Machines are now primarily used to perform the function of drawing-in and tying-in at a high rate. Similar drawing-in of yarns for fill knitting and tufting is also necessary, but the process is not as complex as drawing-in a warp beam. In nonwoven formation, a sliver or a random or plied fiber web is used.

TEXTILE SUBSTRATE FORMATION

Textile substrates are formed from yarns or fiber webs by several techniques including weaving, knitting, tufting, and nonwoven formation. In addition, composites of textile substrates are formed by methods such as adhesive bonding, formation of back coatings on fabric substrates, and flocking. Weaving involves interlacing two sets of yarns usually at right angles to one another using a loom. The warp yarns are fed into the loom and filling (weft) yarns inserted into the warp using a shuttle or an alternative insertion technique. Knitting involves interconnecting yarns by looping them around one another. In warp knitting the yarns in a warp beam are looped over adjacent yarns in a zig zag repeating pattern to form a fabric, while in fill (weft) knitting a fill yarn is formed into a series of loops that are passed through the loops previously formed in the fill direction. In tufting, yarns threaded through needles are punched through a backing fabric and the loops thus formed are held in place as the needles are withdrawn from the backing followed by formation of the next tuft in the same manner. In nonwoven formation, a fiber web or yarns are entangled or bonded to adjacent fibers through use of mechanical or chemical bonding techniques to make a continuous interconnected web. Composites of textile substrates are formed by bonding two fabrics together by use of an adhesive to form a bonded substrate or backed substrate or by application of cut fibers to an adhesive-coated substrate to form a flocked substrate.

WEAVING

Weaving has been traditionally conducted on looms using a shuttle carrying a package (pirn) containing fill (weft) yarn which inserts the fill yarn into the warp which has been drawn-in and tied-in to the loom. In recent years, many shuttle looms have been replaced with shuttleless systems particularly for simple fabric constructions. The shuttle loom continues to be the most versatile weaving machine capable of weaving the widest range of yarns into fabric. The basic components of a loom are presented in Figure 15-1.

The loom functions in the following manner. The warp beam is connected to a let-off mechanism that meters the warp yarns off the beam as fill insertion proceeds. Each yarn in the warp passes through metal warp stop mechanisms that can detect broken warp yarns, through the eyes of heddles that are contained in the various harnesses used to lift the warp yarns, through the reed used to beat-up the filling yarn, and finally onto

the take-up beam. Each warp yarn passes only through the eyes of heddles
within harnesses that will be used to raise that particular warp yarn. The
more harnesses that are used in raising and lowering the yarns, the more
complex the weave that is possible. The pattern of the warp is determined
by which of the harnesses each of the warp yarns is passed. The actual
raising and lowering of the harnesses within the loom is referred to as
shedding, and the space between the separated warp yarns is called the
shed. The harnesses are raised and lowered by use of cams or a dobby
attachment or can also be raised and lowered individually by use of a
Jacquard mechanism.

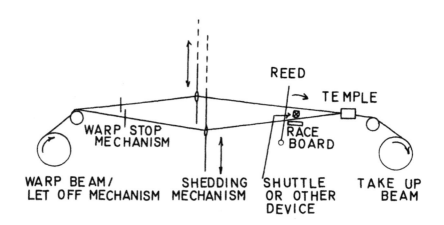

Figure 15-1. Basic components of a loom.

The fill yarn is inserted at right angles to and through the shed by
use of a shuttle or an alternative insertion mechanism by a process called
picking. Below the insertion mechanism is the race board that helps provide support for the shed and the fill insertion area. The reed is placed
between the open shed and the fill insertion area of the shed. After the

fill yarn has been inserted, the reed is used to push the fill yarn tightly into place in the fabric by a process called beat-up. The temple provides uniform tension on the formed fabric to prevent loosening of the inserted fill yarns, and a take-up mechanism keeps proper tension on the fabric as it is formed and taken up. Therefore, the basic repeating sequence of actions in the loom is shedding to open the warp, picking to insert the fill yarn between the separated warp, and beat-up to push the fill tightly into place. The sequence must be carefully timed and synchronized to assure proper insertion. The ratio of fill insertion in shuttle loom varies from 75 to 300 picks per minute, while shuttleless insertion systems operate at 150 to 600 picks per minute. Fabrics woven from staple and filament yarns are found in Figures 15-2 and 15-3, respectively.

Figure 15-2. Woven staple fabric. x50.

Figure 15-3. Woven filament fabric. x45.

Shedding Mechanisms

The cam system is most limiting in the complexity of weaves possible since only 8 to 10 harnesses can be effectively raised and lowered by this method. The cams are generally positioned below the harnesses and raise and lower the harnesses by use of mechanical tappets. For more complex weaves the dobby mechanism or the Jacquard system must be used.

The dobby mechanism is actually mounted on the side of the loom and is capable of raising as many as 20 to 28 harnesses. The dobby mechanism uses a slotted drum and continuous pattern chain containing patterns of pegs on each bar of the pattern chain to lift selected harnesses during shedding.

As a bar of the pattern chain is presented to the slotted drum, the pegs present on that bar enter and block the corresponding slots in the drum. In turn, this prevents hooks attached to the harnesses from entering the slots, engaging, and raising that harness. Where the slots are not occupied by pegs, the hooks attached to individual harnesses enter the slots, engage, and that harness is raised. When the next bar on the pattern chain is presented to the slotted drum, a new series of harnesses are raised depending on the pattern of pegs on that bar.

In the Jacquard system, each warp yarn is attached by an individual heddle to a draw string mounted above the loom and may be raised and lowered independently. The draw strings from the warp yarns pass parallel to each other through separate holes in a perforated board. A series of horizontal needles mounted above the perforated board are pressed against a continuous series of cards having a pattern of perforations. Each cord is in turn attached to a rod that passes through the eye of one of these needles. The upper ends of these rods are hooked and rest over a series of bars called griffes. Where the card has a perforation the needle passes through the perforation, the hook for that warp remains on the bar, and as the bar is raised the warp yarn is lifted. Where no perforation is present, the hook is disengaged and the warp yarn is not lifted. By this method extremely complex and intricate patterns can be developed. Due to the complexity of the mechanism, the weaving rate is much slower with Jacquard than with dobby or cam mechanisms.

Fill Insertion

Until recent years fill (weft) insertion was carried out by traverse of a shuttle containing a package of fill (pirn) within the shed back and forth across the width of the warp. As a result, fabrics produced by shuttle weaving have a selvedge (edge) in which the fill turns in a U at the edge of the fabric to return as the next row of fill in the fabric. Different filling yarns can be inserted by use of multiple shuttles in a magazine arrangement. Because movement of the shuttle back and forth across the fabric is necessary in a shuttle loom, a mechanism for projection (picking) and checking of the shuttle at both sides of the loom is required. The picking stick strikes the shuttle to provide the force necessary to accelerate the shuttle to sufficient velocity to rapidly travel across the width of the loom (picking). As the shuttle nears the other side of the loom, damping mechanisms slow or decelerate and ultimately stop the shuttle (checking), so that it is ready for rapid return across the loom width. The picking and checking action of a conventional loom requires large amounts of energy, and a high degree of vibration is inherent in shuttle systems. In order to minimize energy consumption and machine vibration and to increase the rate of fill insertion, a number of shuttleless systems of fill insertion have been developed. Although these systems are not as versatile as shuttle fill insertion, the improved efficiency and reduced noise levels of these insertion methods make them quite suitable for weaving of less complex weaves. Ultimately, shuttleless systems are expected to be used in production of 80% of all woven fabric.

The shuttleless systems can be divided into two major categories, mechanical systems and fluid systems. Mechanical systems include the use of grippers or rapiers (single and double), while fluid systems use an air jet or water jet. The major fill insert methods are presented in Figure 15-4. The shuttleless systems all insert individual premeasured lengths of fill yarn. Therefore a fill package does not have to be carried across the shed thereby greatly reducing the energy required for fill insertion. The premeasured fill is generally only introduced from one side of the loom, and a traditional stable selvedge with the fill yarn turning back on itself is not produced. The edge of the fabric produced by the shuttleless systems is normally fringed. However, the fringed selvedge can be reinforced by use of a higher density of warp yarns at the selvedge or by attachments producing tucking-in of the selvedge or a leno selvedge. In shuttleless looms, the fill yarn is usually premeasured and cut and held in place by a vacuum tube or on a storage drum prior to insertion.

The gripper system is most closely related to the shuttle system. The gripper is fired as a projectile across the width of the loom carrying a single length of fill. The gripper is much smaller and lighter than a shuttle, since it does not need to carry a fill package. Multiple grippers are used, and a gripper after insertion of fill yarn is returned by a conveyor system back across the loom. A multiphase gripper system is also used in which a series of grippers each carrying fill yarns are conveyed across that loom in sequence by use of a magnetic or mechanical drive mechanism. This method must use phased shedding and beat-up motions to permit simultaneous movement of several grippers across the face of the loom. Although the velocity of the multiple grippers across the loom is much slower than use of a single gripper, the composite rate of fill insertion is much faster than more conventional looms.

In the rapier systems, the fill yarn is carried across the warp by a single or two mechanical arms. The rapiers must be removed from the shed prior to beat-up. In the single rapier system the end of the arm contains a clip to hold the fill yarn that releases the yarn after the fill yarn is completely inserted. In the double rapier system, one arm equipped with a clip (giver) conveys the yarn to the middle of the shed, and the taker on the other arm simultaneously is inserted from the other side and takes the yarn across the rest of the shed. Since rigid rapiers effectively double the width of the loom, flexible rapiers that uncoil on fill insertion have been developed that reduce the loom width.

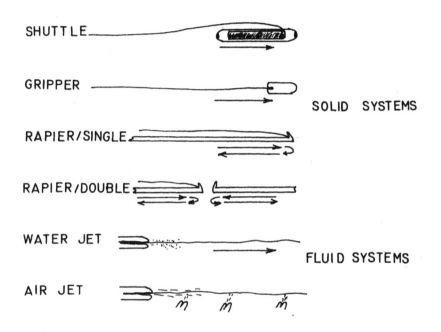

Figure 15-4. Fill insertion methods in weaving.

Fluid fill insertion systems do not use a device to carry the fill yarn across the loom. They operate by impinging sonic velocity water or air jets onto the end of the yarn which accelerates and carries the yarn across the loom. Since liquid water is more cohesive than air and the energy conferred to the water is not as readily dissipated as the energy conferred to air, the water jet is capable of conveying the fill yarn greater distances than an air jet. The major disadvantage of water jet fill insertion methods is related to the hydrophilic character of the water and its ability to dissolve many sizes and to wet out hydrophilic fibers. Therefore, water jets can only be effectively used on hydrophobic fibers such as polyester that are unsized filaments or that contain sizings unaffected by water. To enhance the projection distance of air jet systems, guides are mounted across the loom that are inserted through the warp during fill insertion to provide a turbulence-free path across the loom. Also booster jets are often mounted periodically across the loom and fired sequentially as the fill yarn is inserted to assist in carrying the yarn across the width of the loom.

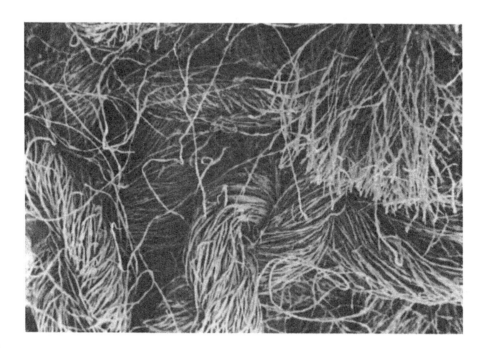

Figure 15-5. Loop pile woven staple fabric. x30.

Special Weaving Methods

When pile type fabrics are produced on a loom an additional warp is necessary. The tension on one of the warps is released before beat-up permitting warp loops to be formed. Also, two warps can be used to weave a double fabric with yarns connecting the two fabric sections. When these yarns are cut, two cut pile fabric pieces are then formed. A looped pile and sheared pile-woven fabric are found in Figures 15-5 and 15-6. Carpets can be woven on looms using more than one warp and a complex fill insertion system. The loops formed in the warp direction are held in place by wires inserted through them from the sides of the loom until the loop is firmly locked in place by the weaving of the backing and then removed. Carpet looms are very complex and carpet formation is very slow. Therefore, most carpeting is produced by tufting.

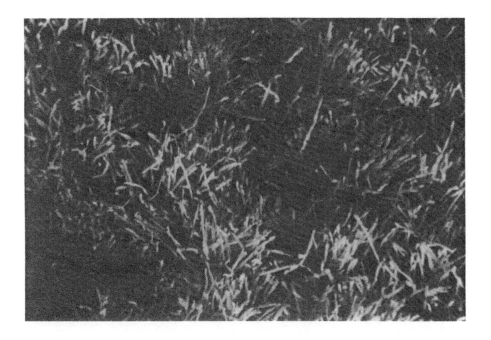

Figure 15-6. Sheared pile woven staple fabric. x40.

Very stable fabric structures can be produced by bringing two warp yarns into the loom at 60° angles to one another and inserting fill in the normal fashion to form a triaxial fabric.

KNITTING

Knitted fabrics are formed by use of hooked needles to interconnect loops of yarn to form a pattern. The needles used are one of three types, bearded, latch, or compound. The hooked needles pull the yarn through a previously formed loop and must close as they pass through the loop and use different closing mechanisms depending on needle type. The bearded needle uses a presser bar to close the needle, while the latch on a latch needle is closed by the loop of yarn it passes through. Compound needles are opened and closed by a programmed mechanism such as a cam mounted near the base (butt) of the needle. These types of needles are used for both warp and fill knitting. In warp and fill knitted structures the row of loops running horizontally across the fabric are referred to as courses, whereas the rows of loops running vertically up the fabric are called wales. Examples of a warp and fill knit are found in Figures 15-7 and 15-8.

Figure 15-7. Warp knit. x30.

Figure 15-8. Fill knit. x30.

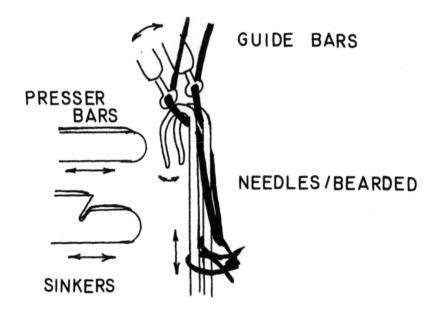

Figure 15-9. Basic elements of a warp knitting machine.

Warp Knitting

Warp knitting is the knitting method most related to weaving. Warp knitting uses a warp beam prepared in the same way as for weaving. The warp is fed into the warp knitting machine, and finished warp fabric is taken up in much the same fashion found on a loom. The actual mechanism of fabric formation is quite different, however. The basic elements and motions of one element of a warp knitting machine are found in Figure 15-9. In a warp knitting machine, there are one or more warp yarns per needle in the machine and each warp yarn is attached to and passes through the eye of an individual guide bar that is used to position the yarn for knitting and to move the yarn from needle to needle to give the characteristic zig-zag pattern of loops running vertically up the warp knitted fabric. The

needles are either bearded, latched or compound and individually mounted
vertically in a horizontal row attached to a straight bar. The guide bars
are mounted in front of the horizontal row of needles and move back and
forth in concert in predetermined patterns to present the yarns to the
appropriate needles for knitting. The more warp yarns per available
needle, the more complex a knitted structure that is possible. In forming
fabric the guide bars wrap the warp yarns across the faces of the individual needles. The needles then close by the mechanism appropriate to that
type needle and the yarns are drawn down through the loops formed in the
previous cycle. The needles then move back up into their original position
and open to receive the next yarn or yarns. Sinkers are used to hold the
loops taut and in place, while the guide bars introduce the yarns to the
needles. In general, the simplest warp knitting machines use bearded
needles and limited guide bars per needle to form tricot fabrics. In these
machines a presser bar is mounted parallel to the needle bar to close the
needles as the warp yarns are drawn through the set of loops formed in the
previous cycle. The more complex Raschel, Milanese and Simplex machines
are capable of more extensive patterning and use latched or compound
needles. The production speed and efficiency of warp knitting machines is
very high and represents the fastest means of converting filament yarn into
fabric. Warp knitted fabrics can be stabilized by introduction of warp
and/or fill yarns into the warp knit structure during knitting. Pile or
terry fabrics are produced by introduction of tow into the knitted structure followed by cutting to form a pile or by use of multiple warp yarns to
introduce terry loops, respectively.

Fill (Weft) Knitting

Fill (weft) knitting involves insertion of a filling yarn into the
knitted structure by drawing the fill yarn as a successive row of loops
through the previously knitted loop structure to form a new course of loops
running across the knitted structure. The yarn may be fed back and forth
across the knitted fabric to form a flat fabric or successive yarns may be
laid in the same direction spaced behind each other to form a tubular fabric. The basic elements of a segment of a fill knitting machine are presented in Figure 15-10. In fill knitting the needles operate in a
sequenced sine-shaped wave. After the needle takes the yarn, it starts
down through the previously made loop and the needle closes by the process
appropriate to the needle type. After the needle with yarn passes completely through the previous loop, the needle opens, releases the newly
made loop and moves upward through the loop to begin the process again. It
is possible to program individual needles within the fill knitter to go

through the full knitting cycle, or go through a partial knitting cycle in which the knitting needle does not fully clear the previous loop (a tuck stitch), or does not go through the knitting cycle at all (float or miss stitch). Such programming is carried out by use of cams or pattern wheels that interact with patterns of tabs or indentations on the butt of the knitting needle. More complex programming of patterns is possible when multiple needle systems are used in conjuction with slot, pin and blade, multiple disk, or punched card and tape mechanisms referred to as Jacquard mechanisms. Multiple needle systems are mounted in parallel rows 60 to 90° out of plane to one another. Such multiple systems also can be used without complex Jacquard programming to simultaneously form double sets of loops in knitting to form ribbed and interlock knitted fabrics depending on the gauging of the needles in relation to each other. Fashioning or shaping of the knitted tube formed is also possible by programming of the number and which individual needles are used within each row of courses formed or through movement of loops from one needle to another before the next row of knitted loops are formed. As in the case of warp knitting, fill knitting can be used to form tufted or pile fabrics or additional warp or fill yarns can be laid into the structure to provide additional stability.

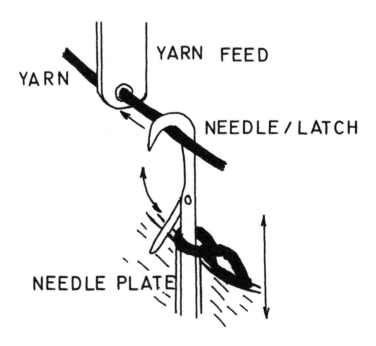

Figure 15-10. Basic elements of a fill knitting machine.

TUFTING AND PILE FORMATION

Tufting is a rapid process for formation of continuous rows of yarn loops across the face of a textile backing substrate. The yarn loops on the tufted substrate then can be cut or sheared to form a cut pile if desired. As mentioned earlier, woven and knitted pile substrates can be formed by introduction of additional yarns into woven and knitted structures. Pile substrates can also be formed by nonwoven and composite forming processes and are discussed below. Because of its high efficiency, today 95% of carpets are formed by the tufting process.

In tufting, a woven or film primary backing material is passed under a row of tufting needles each carrying a yarn. The needles penetrate the backing to form a loop which is held in place as the needle is withdrawn through the backing. Subsequent loops are made in the same manner to form continuous rows of tufted loops on the substrate. After the loops are formed all or some of the tufted loops may be cut by special cutting devices. After the tufted yarn has been inserted into the backing, a layer of adhesive must be applied to the back side of the tufted structure to mechanically fix the tufted yarns in place. Often a secondary backing in the form of a textile substrate is placed and fixed by adhesive over the back side of the yarns or a rubber coating or a rubber or polyurethane foam backing is applied to the back of the tufted substrate. The basic elements of the tufting process are found in Figure 15-11.

The primary and secondary backings used for carpets until recently were woven jute; however, polyolefin and polyester films and coarse woven polyolefin slit film filaments have been increasingly used as backing materials. Nylon yarns account for over 70% of the yarns used to form the tufted face of the substrate, with polyester, polypropylene, acrylic, modacrylic, and wool yarns being used to lesser extents. Nylon dominates the tufted carpet market due to its overall toughness and resiliency. Tufting leads to a wide range of looped pile and cut pile carpet substrates with the number of tufts per unit area, the length of loops or pile, and the fiber type affecting the nature and performance of the carpet.

The tufting process utilizes a large quantity of yarn that is not seen in the finished product, namely the segments of yarn on the backside of the primary backing. Several alternative processes have been developed to minimize this problem, although tufting remains the primary method used. In the alternative processes either premeasured lengths of yarn are forced through a tightly woven fabric backing and then fixed with adhesive or adhesive is applied to the face of the backing substrate and pile is fixed into the adhesive layer on the face of the substrate. Such techniques are limited to the production of cut pile substrates. Flocking is a related method for forming a pile or fuzzy surface on a fabric substrate. In flocking, adhesive is applied to a fabric substrate and then cut fibers or yarns are fixed to the adhesive using mechanical or electrostatic methods to orient the pile vertical to the fabric substrate.

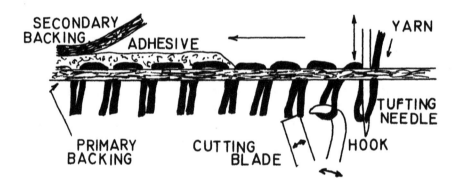

Figure 15-11. Basic elements of a tufting machine.

NONWOVEN FORMATION

Nonwoven textile substrates can be formed through entanglement and/or bonding of fibers in the form of webs or yarns by various chemical and mechanical means. Such methods used to form nonwovens minimize greatly the number of steps required to go from the fiber to the finished substrate and greatly reduce the cost of production of the substrate. The strength, flexibility, and utility of the resulting nonwoven substrate is generally less than that of a similar woven or knitted substrate. Therefore, non-woven fabrics have been used little in apparel other than in interfacing or as felts, but have found extensive use in medical, industrial, and home furnishing applications. Examples of a nonwoven felt and an adhesive-bonded nonwoven are found in Figures 15-12 and 15-13.

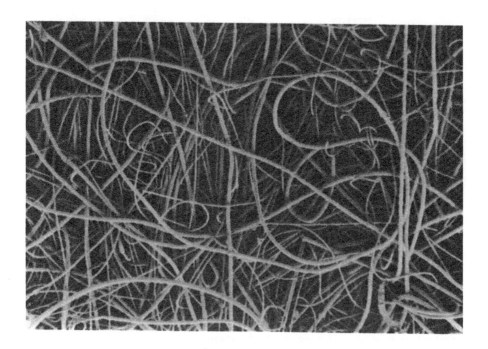

Figure 15-12. Nonwoven felt. x35.

The fiber web or in some cases films or yarns used in nonwoven formation are prepared by conventional processes. Web formation is carried out by carding, garnetting, air laying, or wet laying of staple fibers. The web can also be formed by direct laying of extruded filaments onto a moving belt or through extrusion of a film followed by slitting or embossing and stretching. The webs are laid in a random or oriented fashion with oriented webs being parallel, plated, or crosslaid, in one or more plies. The webs or in selected cases yarns are entangled or are bonded together by four basic processes: mechanical entanglement, stitching, self-bonding, and adhesive bonding.

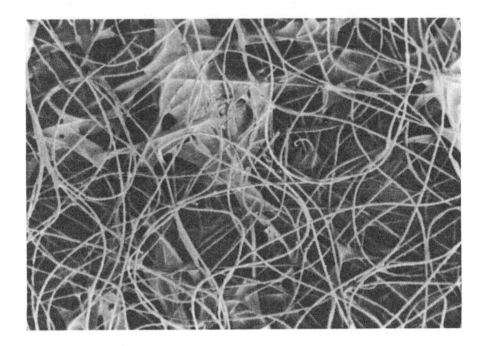

Figure 15-13. Nonwoven adhesive bonded material. x35.

Mechanical Bonding or Entanglement of Nonwovens

A number of processes exist for mechanically entangling or bonding a web to form a nonwoven substrate. One method that is mechanical in nature does not involve a web but rather splitting of a polyolefin film. In this process a film is extruded followed by either a discontinuous pattern cut by knife slitting or by embossing the film followed by biaxial orientation to form a network structure.

Methods that involve mechanical bonding or entangling of fiber web include wet laying, felting, needling, and air or water jet techniques. Wet laying essentially uses the process used for making paper. A slurry of short staple fibers in water or solvent is continuously laid onto a moving wire screen. The water or solvent remaining in the fiber mass that does

not fall through the screen is removed by squeeze rolls and subsequent drying to form a randomly oriented nonwoven substrate. The strength of the
wet laid nonwoven can be increased by adding chemicals or binders to the
slurry during the process. Felting of a fiber web involves the entanglement of fibers in such a way that the fibers "ratchet" on one another and
make the web more dense. This can be accomplished by use of the natural
felting action of scaled keratin fibers such as wool in the presence of
moisture and mechanical action or by heat-induced shrinkage of fibers like
polyester accompanied by mechanical action. Needle punching of a web involves passing the web between bands of needles which continuously punch
through and withdraw from the web. The needles have reverse barbs and
cause reorientation of some fibers in the web and complex entanglement of
the web in the vertical as well as horizontal direction. Needle punching
from both sides of the web will provide a stronger substrate. Water jets
or air jets can also be used to entangle a web. The fiber web is passed
under a series of high velocity water jets or air jets that are programmed
to deliver short bursts of water or air at intervals. The jets disrupt the
web and form bonds through mechanical entanglement of fibers in the web.
In general, mechanical bonding and entanglement methods give soft drapable
fabrics of low to moderate mechanical strength.

Stitching or Stitch Bonding

Yarns or fiber webs are bonded together by stitching the webs together
by a series of needles that operate very similarly to conventional warp
knitting machines. Therefore guide bars must be used as part of the mechanism to interconnect the threads being sewed into the yarns or web being
fed into the machine. Stitch-bonded fabric substrates are usually made
using filament thread to form a reasonably strong but quite flexible structure. Yarns and webs can also be crosslaid in these stitch-bonded processes. The processes operate at very high operating speeds using inexpensive yarns or webs with fabric substrate being formed at 50 to 200 meters/
hr.

Stitch bonding of yarns is usually carried out on Malimo machines,
whereas web stitching is carried out on Maliwatt or Arachne machines. Pile
stitching machines such as Malipol, Araloop, and Locstitch have also been
developed for formation of pile fabrics.

Self Bonding

Techniques have been developed to bond thermoplastic fibers to each other by use of heat and/or solvent. In some cases, low levels of binder are added to the substrate prior to bond formation to assure a higher level of bonding. Spunbonding is the most widely used of these techniques. In spunbonding, a web of continuous filament thermoplastic fibers, such as polyester and polyolefin are extruded and randomly laid onto a moving conveyor belt and then subjected to heat and pressure between embossed rolls to form periodic bonds between fibers in the web. The flexibility of the web formed by spunbonding is determined by both the density of the web and the number of bonding points per unit area. Melding involves use of specially prepared web of sheath/core filaments that can be subsequently bonded by use of heat. The sheath/core fibers have a lower melting polymer substrate making up the sheath than contained in the core. Passing the web between embossed rolls under pressure held at a temperature above the softening point of the sheath polymer, causes the fibers in the web to be bonded at crossover points in the structure. Since the cores of the fibers are unaffected by the treatment, the resulting bonded web is quite strong. A third method, called gel bonding, involves use of a solvent applied to the fiber web. Solvent is metered onto the fiber web causing softening and gelation of the fiber web, and the web is then passed through embossed pressure rollers. The gelled surfaces of the fibers merge at crossover points and on solvent removal by heating form a bond between fibers. The above processes form fabric substrates of moderate stiffness.

Adhesive Bonding

Application of adhesive to the fiber web followed by pressure and heating to cure the binder and bond the fibers in the web to one another is another extensively used method of nonwoven formation particularly for non-thermoplastic fibers such as cellulosics. The adhesive is applied as a solution, suspension, or emulsion to the fiber web by padding, spraying, or printing or by application as a foam. The adhesives used are usually thermosetting or thermoplastic (hot melt) adhesives. After application the web is subjected to heat to drive off solvent (usually water) and to melt and soften a thermoplastic adhesive or to cause reaction or curing of a thermoset adhesive. The adhesive, on curing, tends to concentrate at the fiber crossover points due to capillary action. Nonwoven formation by adhesive bonding tends to form moderate to stiff substrates with the degree of stiffness being dependent on the nature and density of the fiber web, the type and concentration of adhesive used, and the number of bonds per unit area.

COMPOSITE FORMATION

Flexible composite textile substrates are formed by either coating textile substrates with a continuous polymer layer or lamination of two or more textile substrates together by use of an adhesive polymer layer. The polymer coating can be applied neat, from solution, as an emulsion, or in the form of a film or thin foam. Inflexible fiber-polymer composite substrates are formed by imbedding fibers, fiber webs, fiber tows, or fabrics in a stiff polymer matrix. Inflexible composites have found extensive use in engineering and aerospace applications particularly where high performance properties are important. Owing to their inflexibility, they lose any properties characteristic of textile substrates.

In order to get a good bond between the fabric substrate and the polymer coating in laminated and coated fabrics it is essential to get a good adhesive bond between the textile substrate and the coating or adhesive. This is particularly important since the substrate and polymer may have very different stretch and recovery properties. Adhesive failure under stress is the most probable cause of failure in these materials. In coated fabrics, elastomeric properties are necessary to achieve a serviceable coated substrate.

In coated substrates, the polymer coating is metered and spread evenly onto the fabric surface as a controlled viscosity solution, aqueous emulsion, or hot melt polymer followed by drying and curing if needed. A urethane foam can be bonded to a textile substrate, by carefully melting the foam surface with a flame and joining the melted foam surface to the textile substrate under pressure followed by cooling. Laminated fabrics are bonded under similar conditions except that two fabric faces are brought together and bonded by the polymer layer used.

IV. Preparation, Dyeing, and Finishing Processes

16. Preparation and Drying

Before dyeing and finishing a textile substrate, it (griege or gray goods) must undergo preparation to assure complete wetting, uniform application of dye or finish, and/or to remove color or yellowness from the substrate that would interfere with coloration and the visual aesthetics of the textile. Fabric preparation includes wetting out of the fabric, scouring to remove natural waxes or spin finishes on the substrate, desizing to remove sizing from woven and warp knitted fabric, and bleaching to remove color (yellowness) from and to brighten the fabric. Natural fibers in the form of top, sliver, or yarn often undergo a degree of preparation prior to substrate formation. Following fabric preparation and dyeing or finishing, the textile substrate must be dried. Drying processes are currently under intense examination, since they account for nearly two thirds of the energy consumed in textile wet processing.

PREPARATION

In order to conduct the processes associated with preparation, the textile substrate must effectively be wet out. Surfactants or surface active agents are used to lower the surface tension of water in relation to the fiber substrate so the water will wet the substrate. Surfactants are organic molecules that effectively bridge the transition between water and the more hydrophilic fiber surface through combination of a polar head and a hydrocarbon tail within a single molecule.

Surfactants are also necessary for scouring processes in which surface waxes or spinning oils must be lifted from the fiber surface and suspended as micelles of oil surrounded by surfactant, followed by rinsing. Scouring processes are necessary to ensure complete removal of oil or wax from textiles so that dyes and finishes can uniformly penetrate the fiber and fabric structure. Low concentrations of sodium hydroxide in addition to detergent are used to scour cotton fabrics to assist in removal of their natural wax coating (Kiering). Scouring of wool is carried out in the fiber state prior to carding and spinning and is treated in Chapter 14. Sometimes additional scouring of wool substrates may be necessary to remove excess spin finishes or residual lanolin prior to dyeing and/or finishing. Use of solvent-based scouring techniques for synthetic fiber substrates has become more popular in recent years.

Sizing applied to warp yarns prior to weaving or warp knitting by the slashing process must be removed prior to dyeing or finishing to assure even and uniform application of the dye or finish. Sizes are not as readily removed as spin finishes and spinning oils, so special techniques must be used. Sizes can consist of starches, modified starches, and adhesives based on synthetic organic polymers. Starches and modified starches generally may be removed by dilute acid or enzyme treatment, whereas synthetic adhesive sizes can be removed by specialized short washing treatments.

Many natural fibers and some synthetic fibers possess a degree of yellow coloration and must be bleached to brighten and remove yellow coloration from the textile substrate prior to dyeing and/or finishing. Since bleaching is a chemical finishing technique, it is discussed in Chapter 18 under Optical Finishes.

DRYING

After textile processes including sizing and desizing, scouring, dyeing, and finishing, the textile fiber substrate must be dried before further processing can take place or before the substrate can be delivered as the final product.

Wet textile substrates can be considered an irregular but somewhat ordered array of interlocking fibers making up a wet porous fibrous structure. Water within the structure lays between individual fibers being held by capillary action or by cohesive interactions between water molecules. Moisture is also present within the fiber depending on the moisture regain

characteristics of that fiber. In textile processes water not firmly held
by capillary action or water within the fiber is removed by passing the
substrate between squeeze rolls or by passing it over a narrow slit from
which sonic velocity steam blows water from the fabric. Such processes can
reduce the water uptake of the fabric to approximately 30 to 120% wet pick-
up with sonic steam giving the lowest values. The water remaining then
must be removed by convection, conduction, or radiation heating.

Drying of textile substrates is carried out most often by convection
heating. The textile fabric under tension is passed into a convection oven
heated to a temperature well above the boiling point of water. Hot air
passing over the wet textile evaporates water from the fabric surface, and
internal water moves by capillary and related actions to the fabric surface
and in turn is evaporated. The resulting moist air is continuously removed
as dry heated air is introduced into the oven. The tension devices holding
the edge of the fabric remove wrinkles and introduce a uniform width to and
tension on the fabric. These drying units are referred to as stenter or
tenter frame dryers.

Conduction heating finds more limited use in drying of textiles and is
used more in high temperature curing of finishes on textile substrates. In
conduction heating, the textile fabric is brought in contact with and
passed through a set of heated rolls. Since the fabric is in direct contact with the heated metal rolls, heat is transferred more quickly to the
textile than possible by convection heating. The basic process of drying
remains the same for both processes with heating of the outside of the textile structure, and drying from the outside to the interior of the structure.

In radiation heating, the wetted textile structure is irradiated by
thermal radiation using a heat source such as an infrared lamp. On striking the wetted surface the radiation is absorbed and the surface is heated.
Radiation heating has limited use for drying textile structures and is used
more often in curing or pre-drying printed textiles.

Radio frequency and microwave energy can be used to heat wetted textile structures. The textile structure can be heated uniformly throughout,
since water as a polar medium absorbs the energy. Radio frequency and
microwave heating is efficient for initial drying, but as the moisture
level decreases, there is less water to absorb the energy and the process
is not as viable. Therefore, these processes are used in pre-drying fabrics and heating of wet packages.

In recent years the efficiency of drying textiles has been improved through insulation of the heating units and through installation of heat exchangers to remove heat from the outflow of warm moist air from the heating units.

17. Color, Dyes, Dyeing and Printing

COLOR THEORY

Color is defined as the net response of an observer to visual physical phenomena involving visible radiant energy of varying intensities over the wavelength range 400 to 700 nanometers (nm). The net color seen by the observer is dependent on integration of three factors: (1) the nature of the light source, (2) the light absorption properties of the object observed, and (3) the response of the eye to the light reflected from the object. The relative intensities of the various wavelengths of visible light observed by the eye are translated by the mind of the observer resulting in the perception of color. In color measurement, the human eye is replaced by a photocell which detects the light energy present at various visible wavelengths.

Visible light is a narrow band of electromagnetic radiation from 400 to 700 nm (1 nm equals 10^{-9} meters) detected by the human eye. Radiation falling below 400 nm is ultraviolet radiation, and that falling above 700 nm is infrared radiation; both are unseen by the human eye. If pure light of a given wavelength is observed, it will have a color corresponding to that wavelength. Pure wavelengths of light are seen when white light is refracted by a prism into a "rainbow" spectrum of continuous color. Light sources such as sunlight, incandescent light, and fluorescent light are continuums of various wavelengths of light with the relative amounts of the various wavelengths of light being dependent on the overall intensity and type of light source. Sunlight at noon has very nearly the same intensity of each wavelength of light throughout the visible spectrum, whereas at dusk sunlight is of lower intensity and has greater quantities of the longer, red wavelengths than of shorter, blue wavelengths. Fluorescent lights generally contain large amounts of shorter, blue wavelengths, while incandescent tungsten lights contain a large component of longer, red wavelengths compared to noon sunlight. Differences in intensity and wavelength distribution between light sources has a profound effect on the color observed for a dyed textile, since the textile can absorb and reflect only that light available to it from the source. When a dyed fabric appears different in color or shade under two different light sources, the phenomenon is referred to as "flare." When two fabrics dyed with different dyes or dye combinations match under one light source but not under another, the effect is called "metamerism."

When light from a source strikes a dyed textile surface, different portions of the light of the various wavelengths are absorbed by the dye, depending of the structure and light absorption characteristics of the dye. Light not absorbed by the dye on the textile is reflected from the surface as diffuse light, and the observer sees the colors shown in Table 17-1. The color seen is a composite of all the wavelengths reflected from the fabric. If significant direct reflectance of light from the fabric occurs, the fabric exhibits a degree of a gloss. If little or no light throughout the visible range is absorbed by the fabric and the majority of light is reflected, the fabric appears white. If the fabric absorbs all of the light striking it, the fabric is black. If uniform light absorption and reflectance across the visible wavelengths occurs at some intermediate level, the fabric will be a shade of grey.

Table 17-1. Colors After Absorption/Reflectance

Wavelength of Light Absorbed (nm)	Light Absorbed by Dyed Textile	Color Seen by the Observer
400-435	Violet	Yellow-green
435-480	Blue	Yellow
480-490	Green-blue	Orange
490-500	Blue-green	Red
500-560	Green	Purple
560-580	Yellow-green	Violet
580-595	Yellow	Blue
595-605	Orange	Green-blue
605-700	Red	Blue-green

The dye absorbs discrete packages or quanta of light, and the dye molecule is excited to a higher energy state. This energy is normally harmlessly dissipated through increased vibration within the dye molecule as heat, and the dye is then ready to absorb another quantum of light. If the dye cannot effectively dissipate this energy, the dye will undergo chemical attack and color fading or color change will occur or the energy will be transferred to the fiber causing chemical damage.

Organic molecules that contain unsaturated double bonds are capable of absorbing light within a given wavelength range (usually in the ultraviolet). If these double bonds are conjugated and alternate within the molecule, they are able to mutually interact with one another as a cloud of π electrons. If sufficient conjugation exists, the molecule will partially absorb light in the lower energy visible wavelength range and will be considered a dye or a pigment. In general, dyes are colored molecules soluble or dispersible in water or solvent media which can penetrate the fiber on coloration, whereas pigments are not dispersible and must be mechanically entrapped in or locked to the fiber by a binding resin. A series of colorless dyes exist which are called "fluorescent brighteners." These dyes absorb little light in the visible region, but absorb radiation at unseen ultraviolet wavelengths and then emit this radiation at long blue wavelengths to provide an optical bleaching effect on fabrics. Fluorescent dyes combine fluorescence with visible light absorption characteristics to give extremely bright colors, since unseen ultraviolet light is also being made visible to the eye.

The unsaturated groups which can be conjugated to make the molecule colored are referred to as "chromophores." Groups which enhance or alter the color within a conjugated system through alteration of the electron density are referred to as "auxochromes." A series of typical chromophores and auxochromes follows:

CHROMOPHORES — ⬡ , =⬡= $-N\equiv N-$, $-N\stackrel{\stackrel{\textstyle O}{\uparrow}}{\equiv} N-$, $-CH\equiv CH-$

AUXOCHROMES $--CH_3, -OCH_3, -OH, -NH_2, -NO_2, -SO_3Na$

The third component in color is the observer, which can be the human eye or a photodetector in a color instrument. Most color measurement systems are based on a standard light source and a "standard observer" for quantitative measurement of color. The human eye doesn't respond uniformly to color throughout the visible region, but gives maximum response in the middle visible wavelengths. In summary, then, the color observed is a composite of three factors, (1) the light source, (2) the object, and (3) the observer, which may be represented as in Figure 17-1. A change in any one of these three factors will affect the net color observed.

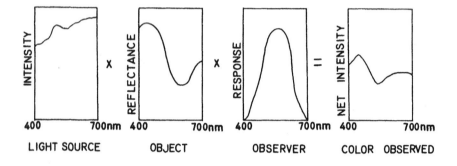

Figure 17-1. Observed color as a composite of three factors.

The color of a textile can be completely defined by the reflectance color spectrum of the fabric as seen above; however, since spectra are of limited value in defining given colors or shades, and numerical color systems have been developed to quantitatively or qualitatively define color. Knowing the relative position of two colors within the color space permits one to determine differences between the two colors. Color is often defined by the following terms: hue, chroma or saturation, and lightness or value. These color terms are defined as follows:

Hue -- basic color type (i.e., red, blue, green, etc.).

Chroma or saturation -- basic difference between color and the closest primary hue (i.e., difference between rose red and true red).

Lightness or value -- relative degree of lightness (i.e., degree of grey).

The Munsell system uses standard hues and numerical values for chroma and lightness to define color.

An array of chips of equal color difference as defined by many observers has been prepared that make up the Munsell Book of Color. The primaries consist of all major hues plus purple arranged logically to form a horizontal circle. The value or lightness makes up the vertical axis varying from 0 to 10 with 0 being black, 10 being white, and the intermediate numbers being varying shades of grey. The chroma is the degree of deviation from the "true" primary and can vary from 0 to 14, with the larger numbers being closest to the "pure" color. To determine the Munsell value for a given color, the Munsell color book is consulted to determine the color designation. For example, if a color had a designation of 5BG 5/8, the sample would be a blue green with a value or lightness of 5 and a chroma of 8. In such a way color can be characterized, but it is not possible to assign strict numerical values to color differences between two dyed samples by use of the Munsell system.

Additive color systems used in color measurement determine the amount of three color primary lights necessary to define a given color using a standard light source and observer. Subtractive systems are used by dyers and colorists, since the net color reflected from a dyebath or dyed fabric is more meaningful in matching dyeings and depends on the amount of the three subtractive primaries present. Color theory and color measurement are complex and are further complicated by an individual's response to physical, physiological, and psychological aspects of color.

Nevertheless, color differences can be effectively measured using additive color systems provided the light source, the observer, and angle of viewing and degree of field observed is defined. In order to clarify and standardize the additive color system and color difference measurement, the Commission Internationale de l'Eclairage (CIE) was formed in 1931. The CIE has provided the definitions and standards necessary for color measurement. The primaries defined by CIE are not real colors, but are imaginary primaries used to define all colors in the color space. The amount of each of these primaries (tristimulus values X,Y,Z) in a given color are used to define the shade and depth of shade for that color. Since the tristimulus values are difficult to plot, the values are normalized and reduced to trichromatic coefficients according to the equations $x = X/X+Y+Z$ and $y = Y/X+Y+Z$. Then the shade of a sample is defined by x and y, and the relative lightness by the Y value which is equivalent to the total reflectance of the dyed textile as observed by the human eye. Color differences (ΔE)

between two samples are determined by the equation $\Delta E = (\Delta x^2 + \Delta y^2 + \Delta Y^2)^{1/2}$. The color space can be mathematically presented in a number of ways and numerous color difference formulas and systems exist.

Recently, the CIE has established a color and color difference system based on rectangular color coordinates called CIELAB. This system uses L rather than Y for lightness, since L is more closely related to the response of the human eye, with black being assigned 0 and white being assigned 100. A and B provide the shade value of the color in rectangular coordinates with A+ being red and A- being green, and with B+ being yellow and B- being blue. Then the equation $\Delta E = (\Delta L^2 + \Delta A^2 + \Delta B^2)^{1/2}$ will give the color difference. Also cylindrical coordinates of L, C, and H°, where L is lightness as defined above, C is the distance from neutral grey shade (C = $A^2 + B^2$), and H° is the arc tangent B/A expressed on a 360° scale where A+ = 0°, B+ = 90°, A- = 180°, B- = 270° may be used if desired.

The concentration of dye on a fabric after a given dyeing time can be determined by three methods: (1) measurement of the decrease in dye concentration in solution with time by ultraviolet-visible spectroscopy, (2) determination of the dye concentration on the fabric dyed for a given time by dye extraction and ultraviolet-visible spectroscopy, or (3) by measurement of the reflectance spectra of the dyed fabric followed by application of the Kubelka-Munk relationship in which $K/S = \frac{(1-R)^2}{2R}$. (K/S is the scattering coefficient and directly related to dye concentration, whereas R is the reflectance of the fabric at a wavelength of maximum absorption).

DYES AND DYE CLASSIFICATION

Dyes as colored unsaturated organic molecules must have affinity for fibers to be effectively applied. The dyes on fibers are physically bound to the fiber by one or more physical forces including hydrogen bonding, van der Waals, or ionic forces and in certain cases chemically bound by covalent bonds.

Dyes may be classified in a number of ways, including color, intended use, trade name, chemical constitution, and basis of application. Of these classification methods, chemical constitution and basis of application have been most widely used. Chemical constitution indicates the major chromophores present in the dye but does not indicate more than such structural aspects of the dye. A classification scheme for dyes has been developed and evolved for use by dyers which is based on the method of application and to a lesser degree on the chemical constitution of the dye class. The classification scheme and major dye classes are outline below.

Dyes Containing Anionic	Dyes Requiring Chemical
Functional Groups	Reaction before Application
Acid dyes	Vat dyes
Direct dyes	Azoic dyes
Mordant dyes	Sulfur dyes
Reactive dyes	
	Special Colorant Classes
	Disperse dyes
Dyes Containing Cationic	Solvent dyes
Groups	Pigments
Basic dyes	Natural dyes

Dyes classified by this scheme are assigned standard designations according to dye class, color, and overall constitution by the Society of Dyers and Colorists in the Color Index (e.g., Acid Blue 141, Vat Green 17, Disperse Red 17).

Dyes Containing Anionic Functional Groups

A number of dyes, including acid, direct, mordant, and reactive dyes, contain functional groups that are sodium salts of sulfonic or carboxylic acids. These functional groups provide water solubility to the dyestuff. The dyes differ in subclassification in their affinity for fibers and/or the presence of special functional groups.

Acid Dyes: The acid dyes are large dyes containing one or more sulfonic or carboxylic acid salt functional groups. These dyes are dyed onto fibers from acid solution, since positive charge development within the fiber in acid solutions acts as a driving force for dye diffusion and migration into the fiber. Only fibers which develop a positive charge in the presence of acid, such as wool, silk, and other protein fibers, nylon, and certain modified synthetics, are readily dyed by acid dyes. Acid dyes on fibers are reasonably colorfast to light and laundering, but mordanting (more complete insolubilization of the dye through reaction with a metal salt) will improve the overall fastness properties of the dye. The color of the dye may be affected somewhat by mordanting; however, pre-metallized acid dyes are a special class of acid dyes which have been reacted with a mordant prior to dyeing and which have sufficient solubility to be dyed under conditions normally used for acid dyes.

ACID BLUE 78

One method of classifying acid dyes involves dividing them into three groups according to their application and wetfastness. Leveling acid dyes dye evenly to a moderate degree of fastness. For wool and protein fibers they are small molecules requiring highly acidic dyebaths for good exhaustion. For nylon, leveling dyes are somewhat larger molecules and are applied from weakly acidic dyebaths. Milling dyes have better washfastness, but generally give duller shades and lack good leveling characteristics. They are larger than leveling dyes and are applied from dilute acetic acid solutions. Acid milling or super milling dyes are applied from neutral solutions and have poor leveling characteristics due to their larger molecular size. They generally are applied to wool and have high wetfastness and lightfastness.

ACID BLUE 158
PRE—METALLIZED ACID DYE

Direct Dyes: Direct or substantive dyes are a special class of dyes which penetrate cellulosic fibers readily and have good affinity for these fibers due to their size and shape. Whereas acid dyes are large and bulky, direct dyes are long, narrow, and flat in molecular structure, which allows them to readily enter the cellulose structure and interact with the cellulose in such a way as to provide good fiber affinity. Direct dyes often contain one or more azo groups connecting aromatic chromophores, thereby providing a straight chain dye molecule. Since charge development is not a primary consideration in diffusion of direct dyes onto cellulosics, the dyes are usually applied from basic solutions in which cellulosics are more stable and more likely to swell.

DIRECT RED 185

Direct dyes fall into three main categories. Class A direct dyes level well even in the presence of high salt concentrations, while Class B direct dyes have poor leveling properties which can be improved by proper adjustment of salt concentration. Class C dyes have poor leveling properties, but exhaust well on raising the temperature of the dyebath in the absence of salt. The direct dyes are reasonably colorfast on cellulosic fabrics, but fastness can be improved by mordanting with metal salts, cationic fixing agents, or formaldehyde or through reaction with diazonium compounds or diazotization of the dye and reaction with a coupling compound.

Mordant Dyes: Mordant dyes are acid dyes that have special sites other than acid salt anion groups that can react with a metal salt mordant. Mordant dyes are "tailor-made" to chelate with metal ions to form a strong organometallic complex of limited solubility and greater colorfastness. The fiber may be dyed initially and then mordanted (postmordanting), dyed and mordanted simultaneously (comordanting), or mordanted and then dyed (premordanting). Of the three methods, postmordanting is preferred. Salts of chromium, aluminum, copper, iron, tin, and cobalt are commonly used as mordants. Since the mordant affects the electron distribution and density within the dye, the color of the dyed fabric tends to change.

MORDANT   BROWN   35

Reactive Dyes: Reactive dyes are dyes which usually have the basic structure of acid, direct, or mordant dyes but which in addition have a reactive
group capable of covalent bond formation with the fiber. Since the fiber
must have reasonable reactivity toward the dye reactive group, application
of these dyes has been limited to cellulosic, protein, and nylon fibers for
the most part.

REACTIVE RED I

The fastness of reactive dyes covalently bound to the fiber is excellent. Reactive functional groups have been selected for incorporation into reactive dyes which will react readily with the fiber after diffusion into the structure but which will not hydrolyze (decompose) in the water solvent used in dye application. Acidic or basic conditions are necessary for successful and rapid reaction of the reactive dye with the fiber, so dye application is carried out at either slightly acid or basic pH (hydrogen ion concentration). Procion dyes are the best known of the reactive dyes.

$$\text{Na O}_3\text{S-D-X} \quad + \quad \begin{Bmatrix} \text{HO-} \\ \\ \text{NH}_2\text{-} \end{Bmatrix} \quad \xrightarrow{\text{H}^+ \text{ or OH}^-} \quad \begin{matrix} \text{Na O}_3\text{S-D-O-} \\ \\ \text{Na O}_3\text{S-D-NH-} \end{matrix}$$

REACTIVE DYE   FIBER   DYE BOUND TO FIBER
D — DYE CHROMOPHORE
X — REACTIVE GROUP

Dyes Containing Cationic Groups (Basic Dyes)

Basic or cationic dyes are colored cationic salts of amine derivatives. Basic dye cations will migrate toward negative charges inside the
fiber. The dyes may be applied to cellulosic, protein, nylon, acrylic, and
specially modified synthetic fibers. Although the dyes generally are of
striking brilliance and intensity, the colorfastness of the dyes on cellulosic, protein, and nylon fibers is generally poor. Colorfastness can be
improved through mordanting with tannins or other complexing agents. Since
the insoluble parent amine is regenerated from basic dyes at basic pH, the
basic dyes are applied from mildly acid or neutral solutions. The dyes are
also important in "one-shot dyeing" of paper products.

$$N(CH_3)_2$$

$$Cl$$

$$C$$

$$+$$

$$Cl^-$$

$$N(CH_3)_2$$

BASIC   BLUE   I

Dyes Requiring Chemical Reaction before Application

Both vat and sulfur dyes must be chemically reduced before application
to a fiber, whereas azoic or naphthol dyes are formed through reaction of
two separate dye components after application to the fiber. These dyes
tend to penetrate the fiber less than other dyes, and care must be exer-
cised in application to get reasonable fastness properties.

Vat Dyes: Vat dyes are usually water-insoluble dyes that can be chemically
reduced in the presence of base to form a water-soluble and colorless leuco
form of the dye, which is then applied to the fiber. Vat dyes can be
readily applied to cellulosic fibers and most synthetic fibers, but care
must be taken in applying the dyes to protein fibers due to the high
basicity of the leuco dye solution which can damage protein fibers.

VAT BLUE I
INDIGO

The dyes are usually indigoids (such as indigo) or anthraquinone derivatives and applied at low (30°-60°) temperatures. After application of the
leuco form of the vat dye, the dye is reoxidized on the fabric by oxygen in
the air or through treatment of the dyed fabric with a mild oxidizing
agent. The vat dyes are reasonably colorfast if poorly held surface dye
has been removed:

$$D \; + \; 2(H) \xrightarrow{\;\;OH^-\;\;} DH_2 \xrightarrow{\;\;(0)\;\;} D$$

| VAT DYE | REDUCING AGENT | LEUCO FORM | OXIDATION | VAT DYE ON FIBER |

Stable sulfate esters of reduced leuco forms of vat dyes are available which do not require prior chemical reduction before application.

Sulfur Dyes: Sulfur dyes are inexpensive complex reaction mixtures of selected aromatic compounds with sodium polysulfide. The sulfur dyes are chemically reduced in the presence of base prior to application to the fiber, and are reoxidized after dyeing on the fiber by oxygen in the air or by application of a mild oxidizing agent such as hydrogen peroxide.

SULFUR GREEN 6

The sulfur dyes are similar in application and fiber affinity to vat dyes, and also are available in a solubilized leuco form:

$$D\text{-}S_X \quad + \quad (H) \xrightarrow{\quad OH^- \quad} D\text{-}S_X H \xrightarrow{\quad (O) \quad} D\text{-}S_X$$

SULFUR DYE   REDUCING   REDUCED OXIDATION SULFUR DYE
           AGENT   DYE   ON FIBER

Sulfur dyes are inexpensive, and have adequate colorfastness properties for most applications.

Azoic Dyes: Azoic or naphthol dyes are formed in situ on the fabric through a coupling reaction of an aromatic alcohol or amine such as naphthol (the coupling component) with a diazonium salt (the diazo component). The dye formed contains an azo group:

DIAZO COUPLING COMPONENT I

AZOIC DIAZO COMPONENT 9

The fabric is usually impregnated first with the aromatic coupling component followed by immersion of the fabric in a solution containing the diazo
component, with the azoic dye being formed instantaneously. Owing to the
instability of the components, the dyeings are carried out near room temperature. The dyes have moderate fastness, and dyeings may be mordanted to
increase fastness.

$$\text{D-NH}_2 \; + \; \text{NO}_2^- \xrightarrow{\text{H}^+} \; \text{D-}\overset{+}{\text{N}}\text{≡N} \xrightarrow{\text{D'H}} \; \text{D-N=N-D'}$$

DIAZO	DIAZONIUM	COUPLING	AZOIC DYE
COMPONENT	SALT	COMPONENT	ON FIBER

Special Colorant Classes

Several types of dyes or colorants do not "fit" logically into the other classifications and have been included in this special classification. Disperse dyes are small polar dye molecules which can be used to dye thermoplastic fibers such as triacetate, nylon, polyester, and other synthetic fibers. Solvent dyes are dyes which are specially formulated so that they can be applied from solvents other than water. Pigments are not dyes at all, but rather colorants that must be incorporated in the fiber during spinning or fixed on the surface of the fiber by use of a polymer adhesive. Natural dyes are complex mixtures derived from natural sources which can be placed in a number of the above classifications.

Disperse Dyes: Disperse dyes were formulated and introduced to permit dyeing of hydrophobic thermoplastic fibers including acetate, triacetate, nylon, polyester, acrylic, and other synthetics. The disperse dyes are small polar molecules, usually containing anthraquinone or azo groups, which do not have charged cationic or anionic groups within the structure. The disperse dyes are sparingly soluble in water and must be dispersed with aid of a surfactant in the dyebath. As the small amount of dissolved disperse dye diffuses into the fiber, additional dye dispersed in solution is dissolved, until the disperse dye is nearly completely exhausted onto the fiber. A special class of reactive disperse dyes has been introduced that can react with fibers like acetate and nylon after diffusion into the fiber. The light- and washfastness of these dyes is generally good, but difficulty has been encountered with fume fading with certain of the disperse dyes. Many disperse dyes have appreciable vapor pressures at elevated temperatures and can be "dyed" onto thermoplastic fibers by sublimation, which involves diffusion of the dye vapors into the fiber.

DISPERSE BLUE 6

Solvent Dyes: Solvent dyes often resemble dyes in other classes, except that these dyes contain groups that improve their solubility in solvents such as alcohols and chlorinated hydrocarbons. It is often possible with solvent dyes to dye certain fibers which would be poorly dyeable from aqueous solution.

SOLVENT YELLOW 3

Pigments: Pigments as a class are colored materials that are insoluble in their medium of application. As a result, they cannot penetrate or become readily fixed to a fiber and must be "locked" onto the fiber surface by use of a polymeric adhesive binder that encapsulates and holds the pigment onto the fibers. Pigments include inorganic salts, insoluble azo or vat dyes, toners, lakes, metallic complexes, and organometallic complexes. Some binders are preformed water-insoluble polymers applied from solvents or as emulsions, whereas others are water-soluble or emulsifiable polymers which can be chemically crosslinked and insolubilized after application by drying and heating. The pigment-binder systems tend to stiffen textiles, and have moderate to poor fastness, since they are surface treatments.

PIGMENT ORANGE 3

Natural Dyes: Natural dyes and pigments are derived from mineral, animal, or plant sources and generally are complex mixtures of materials. A number of natural dyes can be classified as acid or vat dyes and can be readily mordanted. In general, the natural dyes give more muted tones than synthetic dyes and are useful only on natural or regenerated fibers.

NATURAL RED 8
PURPURIN

Dyeing of Blends

Textile structures are often blends of more than one fiber type. Blends containing similar fibers are easy to dye with dyes of similar structure and application characteristics. Care must be taken in dyeing blends containing fibers of highly different dye affinities, if the dyeing is carried out in the same dyebath. The dyes and their auxiliaries must be compatible with one another. When fibers in a blend are dyed the same color, the dyeing is referred to as "union dyeing," whereas dyeing fibers in a blend of different colors is referred to as "cross dyeing." In blends, interesting tone-on-tone, tone-on-white, and differential dyeings are possible by selection of appropriate dyes and dyeing conditions.

APPLICATION METHODS AND FACTORS AFFECTING DYEING

Dyes may be applied to textile structures in a number of ways and at a number of points within the textile construction process. Dyeing or print-ing techniques can be used. Dyeing methods involve application of dye solutions to the textile, whereas printing can be considered a specialized dyeing technique. In printing the concentration of dye is higher, and the dye medium is thick and viscous to limit dye migration on the fabric, per-mitting formation of a design or pattern.

Fiber or stock dyeing involves dyeing of the loose fibers or fiber top or sliver before yarn formation. Colorants can be added to man-made poly-mers before they are formed into fibers to ensure uniform coloration throughout the fiber. This technique is referred to as "solution" or "dope dyeing." Yarns or skeins can be dyed, and piece dyeing of textile fabrics can be carried out. Resist techniques such as tie-dyeing or batik limit dye migration to certain parts of the fabric. Discharge processes use bleaches or other chemicals to remove dye from selected areas of a dyed or printed fabric.

The above processes are used for application of dyes to textile sub-strates in the form of fiber sliver or top, yarn, or fabric by batch (dis-continuous) or continuous methods.

Fiber or stock dyeing of sliver or top is usually conducted in large vats with movement of the dye liquor through the fibers to assure intimate contact of the dye liquor with the fibers. Yarn or skein dyeing is con-ducted by suspending skeins of yarn in an agitated dyebath with possible additional movement of the skeins during dyeing. Yarns can also be wound as packages on perforated spindles or spools and immersed in the dyebath. The dye liquor is then circulated back and forth through the packages. Fabric rolls can also be dyed by this package dye technique. Package dye-ing is often carried out in closed systems at elevated temperatures and pressures. Jig dyeing involves passage of a fabric piece back and forth from one spindle to another through a dyebath, whereas a dye beck contain-ing a winch is used to move a continuous fabric piece through the dye liquor. The above techniques are all batch processes.

The pad-batch method is a specialized batch technique for application of reactive dyes to cellulosic fibers. In the pad-batch method, cellulosic fabric is passed through a concentrated solution of reactive dye, followed by storage of the wet fabric in a vapor tight enclosure for 24 to 48 hours at room temperature, to permit time for diffusion and reaction of the dye with the fabric substrate prior to washing off unreacted dye. The process leads to significant energy savings compared to conventional processes.

Various specialized techniques have been developed for application of disperse dyes to polyester. Unless the dyeing is carried out at 100°C or above, the rate of dyeing is very slow. Dyeing with disperse dyes from aqueous solutions at 120°-130°C to achieve rapid dyeings is common but requires the use of closed high-pressure equipment. Recently, jet dyeing has been introduced, which permits not only high-temperature dyeing but also impingement of the dye onto the moving fabric through use of a venturi jet system. Also, carriers can be introduced to permit dyeing of polyester at atmospheric pressure and below 100°C. Carriers are usually aromatic organic compounds that can be emulsified in water and which have affinity for the polyester. The carriers penetrate the polyester, open up the molecular structure of the fiber (often resulting in swelling of the fiber), and aid in passage of the disperse dye across the dye solution-fiber interface and within the fiber. Suitable carriers include aromatic hydrocarbons such as diphenyl and methylnaphthalene, phenolics such as o- and p-phenylphenol, halogenated aromatics such as the di- and trichloro-benzenes, aromatic esters including methyl salicylate, butyl benzoate, and diethylphthalate, and benzaldehydes. Carriers must be removed after dyeing, and the presence of carriers in mill effluents presents a problem because of their toxicity.

Continuous dyeing is carried out on a dyeing range where fabric or carpet is continuously passed through a dye solution of sufficient length to achieve initial dye penetration. The dye on the fabric or carpet is fixed by subsequent steaming of the substrate. Recently, foamed dye formulations have been applied to fabrics, thereby effectively reducing the dye liquor to fabric ratio and reducing energy and effluent treatment costs.

A novel approach to continuously dyeing polyester with disperse dyes involves sublimation of disperse dye under heat and partial vacuum into polyester by the technique called "thermosol dyeing." Polyester containing disperse dye applied to the fiber surface is heated near 200°C under partial vacuum for a short period of time. At this temperature, the molecular motion within the polyester is high, permitting the dye vapor to penetrate into the fiber. On cooling, the disperse dye is permanently trapped and fixed within the fiber. The above processes are graphically represented in figure 17-3.

Figure 17-3. Dyeing processes.

In printing, the printing paste is applied through use of direct transfer dye using a block or engraved roller or through application using a partially marked flat or rotary screen and squeegee system.

Printing of polyester by disperse dyes can be accomplished by heat transfer printing, which is a modification of thermosol dyeing. In this process disperse dyes are printed onto paper followed by bringing the polyester fabric and printed paper together under pressure with sufficient heating to cause diffusion of disperse dyes into the polyester. Block, flat screen, and heat transfer processes are batch processes, whereas engraved roller and rotary screen printing are continuous processes. Special techniques using dyeing solutions which give printed style fabrics have been developed. In these processes, multiple jets of different dye solutions are sprayed in programmed sequence onto the fabric as it passes under the jets to form patterns with definition very nearly like that of prints. Specialized techniques for formation of patterns on carpets have been developed. In one process dye solution is metered and broken or cut into a pattern of drops which are allowed to drop on a dyed carpet passing underneath to give a diffuse overdyed pattern on the carpet. Representations of the printing and printed style processes are found in Figure 17-4.

Physical factors such as temperature and agitation and auxiliary chemicals added to the dyebath or printing paste can alter the rate of dyeing (dyeing kinetics) and/or the total dye absorbed by the fiber (dyeing thermodynamics). In dyeing, the rate of dyeing of the fiber is dependent on the rate of migration of dye in solution to the fiber surface, the rate of diffusion of dye at the fiber interface, and the rate of diffusion of dye in the fiber matrix. Agitation of the dyebath effectively eliminates the effect of dye diffusion to the fiber in the dyebath. The rate of dye passage across the fiber-liquid interface is rapid in most cases; therefore the rate of dyeing is solely determined by the rate of dye movement within the fiber matrix. As the temperature of dyeing is raised, the rate of "strike" of dye onto the fiber and diffusion in the fiber increases, whereas the total amount of dye present in the fiber at equilibrium decreases. In other words, heating a dyebath speeds dyeing but decreases the total dye exhausted on the fabric. Dyeing is usually carried out at a temperature above the glass transition temperature (T_g) of the fiber, since the molecular segments of the polymers within the fiber have more mobility and permit more rapid dye diffusion above that temperature.

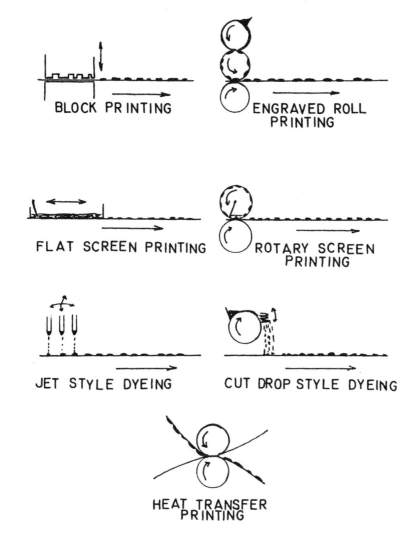

Figure 17-4. Printing and printed style processes.

A number of chemical reagents (auxiliaries) are added to the dyebath to alter in some manner the course of dyeing. Acids or bases may be added to the dyebath to induce charge formation in the fiber in order to increase dye diffusion or to promote reaction of the dye with the fiber as in the case of reactive dyes. In addition, bases may be used to open up the structure of cellulosics to improve dyeing or to aid in dissolving of vat and sulfur dyes. Common inorganic salts such as sodium chloride or sodium sulfate may be added to a dyebath and act as leveling agents to retard the rate of dyeing and give a more even dyeing. Surface-active agents such as anionic detergents can act as leveling agents and also permit the rapid and complete fiber wetting needed to give even dyeing. Water softening chemicals such as sodium hexametaphosphate are added to bind or chelate hardwater calcium or magnesium ions that may interfere with dyeing. Reducing agents are added to vat and sulfur dyes to react with the dye to give the leuco form. Added organic solvents and/or certain chemical reagents may increase the dye solubility and rate of penetration into the fiber. Carriers are aromatic derivatives added in disperse dyeings to open up the fiber structure of polyester and other thermoplastic fibers and to increase the rate of dye diffusion. Urea has a similar effect in speeding dyeing of cellulosic and protein fibers. In printing pastes, natural and synthetic gums and thickeners are necessary to provide viscosity and thickness to the dye paste and to limit dye migration across the fabric face. Heating of fabrics following printing is necessary to achieve good dye penetration into the fiber, and superheated steam is preferred, since wet fixation causes a faster rate of dye migration than dry heating methods.

The fastness of the dyeing will be dependent on the location of the dye within the fiber, chemical and physical forces holding the dye to the fiber, and the stability of the dye-fiber combination to environmental factors. Dye distributed evenly throughout the fiber in small aggregates is preferred, whereas surface or ring dyeing leads to poor washfastness.

DYES APPLIED TO FIBER CLASSES

Dyes for Cellulosic Fibers

Cellulosic fibers can be dyed readily using a number of dye classes. The less crystalline regenerated cellulosics (rayon) dye more readily than the more crystalline natural cellulosic fibers such as cotton. Cellulosics can be dyed or printed using direct, reactive, basic, vat, sulfur, or azoic dyes. Cellulosics can also be colored using pigment-binder systems. Mordanting of direct and basic dyes on cellulosics improves overall fastness. Since the hydroxyl groups in cellulosics are less reactive than many functional groups found in protein fibers, reactive dyes such as the dichlorotriazine (Procion M) or dichloroquinoxaline (Levefix) dyes are often used on cellulosic fibers due to their higher reactivity. The general stability of cellulosic fibers to basic solutions permits them to be effectively dyed by vat, sulfur, and azoic dyes without significant damage.

Dyes for Cellulose Ester Fibers

Acetate and triacetate fibers can be effectively dyed using disperse dyes. The rate of dyeing is more rapid with the more hydrophobic triacetate fibers than with acetate. Under special conditions, azoic and vat dyes may be used to dye these fibers. Acetate fibers also have affinity for selected acid and direct dyes. Since acetate loses its luster above 85°C, dyeings must be carried out at or below this temperature. Addition of pigments or solvent-soluble dyes to the acetate or triacetate spinning "dope" prior to fiber spinning leads to colored fibers possessing excellent colorfastness, although the colors available are limited.

Dyes for Protein Fibers

Protein fibers are the most readily dyed fibers due to the numerous reactive functional groups present. They can be dyed with a wide range of dyes under acid, neutral, or slightly basic conditions. Since the keratin fibers are less crystalline and oriented than secreted fibers such as silk, they tend to dye more rapidly and more readily to deeper shades.

Under acid conditions amino groups in the protein fibers are protonated to form NH_3^+ groups. In this form, they are able to attract dyes containing acid anions including acid, direct, mordant, and reactive dyes. Special premetallized acid dyes of sufficient solubility are used to dye protein fibers to fast colors. The functional groups available in protein fibers are more reactive than hydroxyl groups in cellulosic fibers. Reactive dyes of more limited reactivity have been developed especially for protein fibers. Protein fibers complex very readily with multivalent metal cations. Acid dyes and mordant dyes may be rendered very fast by mordanting with metal salts, and chromium salts are especially effective as mordants. At neutral or slightly acid pH, protein fibers may be dyed with cationic or basic dyes; however, the fastness of the dyed fiber is poor without mordanting with tannic acid or other mordants for cationic dyes. Azoic and vat dyes find only limited use on protein fibers due to the damaging effect that basic solutions of these dyes have on protein fibers. Sulfate esters of reduced vat dyes can be used effectively on protein fibers, however. Natural dyes from many sources have good affinity for protein fibers and are used extensively in the crafts area and often in conjunction with mordants.

Dyes for Polyamide Fibers

With the exception of the aramid fibers, the polyamides dye readily with a wide variety of dyes. Since the polyamides contain both acid carboxylic and basic amino end groups and have a reasonably high moisture regain, the fibers tend to dye like protein fibers such as wool and silk. Since the molecular structure is somewhat more hydrophobic, more regular, and more densely packed in the polyamides than in protein fibers, they also exhibit to some degree the dyeing characteristics of other synthetic fibers such as polyesters and acrylics. Due to their highly regular molecular structure and dense chain packing, the aramid fibers resemble polyester and are dyed only by small dye molecules such as disperse dyes. Polyamides such as nylon 6, 6,6, and Qiana can be readily dyed with dyes containing anionic groups, such as acid, metallized acid, mordant dyes, and reactive dyes and with dyes containing cationic groups such as basic dyes. Acid dyes on nylon can be mordanted effectively for additional fastness; however, the colorfastness of basic dyes is poorer and more difficult to stabilize by mordanting. Vat and azoic dyes can be applied to nylons by modified techniques, and polyamides can be readily dyed by disperse dyes at temperatures above 80°C. Aramids can only be dyed effectively with disperse dyes under rigorous dyeing conditions. The biconstituent fiber of nylon and polyester can be effectively dyed by several dye types due to the nylon component, but for deep dyeings disperse dyes are preferred. Nylon 6 and 6,6 are produced in modifications that are light, medium, or deep dyeable by acid dyes or specially dyeable by cationic dyes.

Dyes for Polyester Fibers

Owing to their high crystallinity and hydrophobicity, the polyester fibers are extremely difficult to dye by normal dyeing techniques unless the fiber has been modified, as in the case of modified terephthalate polyesters. A limited amount of polyester is solution dyed through incorporation of dye or pigment into the polymer melt prior to spinning. It is more common to use this technique to incorporate fluorescent brightening agents into polyester. Only smaller, relatively nonpolar dye molecules can effectively penetrate polyester; therefore disperse dyes have been the dye class of choice for the fiber.

Azoic dyes and pigment-binder systems have also found limited use on polyesters. Polyester modified with appropriate comonomers can be dyed at lower temperatures or with acid or basic dyes depending on the nature of the modifying groups.

Dyes for Acrylic Fibers

The nature and distribution of acrylonitrile and comonomer or comonomers in the acrylic fibers affect the overall dyeability and the classes of dyes that may be used in dyeing these fibers. Both acrylic and modacrylic fibers can be dyed using disperse dyes, with the more hydrophobic and less crystalline modacrylic being more dyeable with this dye class. The polar cyanide groups in the acrylonitrile unit of these fibers have some affinity for acid dyes and particularly mordanted systems containing copper or chromium ions. Addition of an acid or basic comonomer such as acrylic acid or vinyl pyridine as comonomer imparts improved dyeability with basic and acid dyes, respectively, for these fibers. Vat dyes can be used on acrylic fibers to a limited extent.

Dyes for Polyolefin Fibers

Polyolefin fibers are hydrophobic, and the molecular chains within the fiber are tightly packed. Therefore it is extremely difficult to dye polyolefin fibers or to increase their affinity to dyes. Colored inorganic salts or stable organometallic pigments have been added to the polymer melt prior to fiber spinning to color the fibers. Also, nonvolatile acids or bases or materials such as polyethylene oxides or metal salts have been added to the polymer prior to fiber formation to increase the affinity of the fiber for disperse, cationic, acid, or mordant dyes. Polyolefin fibers can be chemically grafted with appropriate monomers after fiber formation to improve their dyeability.

Dyes for Vinyl Fibers

The vinyl fibers, with the exception of vinal and vinyon-vinal matrix fibers, are extremely hydrophobic and difficult to dye, and consequently they can be dyed only through pigmentation of the polymer melt before fiber formation or through dyeing with disperse dyes. Vinal and vinyon-vinal matrix fibers dye readily with dyes used on cellulosics including direct, mordant, reactive, vat, and sulfur dyes.

Dyes for Elastomeric Fibers

Since the elastomeric fibers are often a component in the core of blended yarns, coloration is not important in all applications. Rubber fibers cannot be dyed readily and are colored through mixing of pigments into the rubber prior to extrusion into fibers. Spandex fibers are more dyeable and can be dyed with acid, reactive, basic, or vat dyes. Anidex can be dyed with disperse or basic dyes. The nylon component of spandex-nylon fibers can readily be dyed with acid, basic, disperse, or vat dyes.

Dyes for Mineral and Metallic Fibers

The mineral and metallic fibers are essentially undyeable, and special techniques must be used to impart color to the fibers. Thermally stable ceramic pigments can be added to molten glass prior to fiber formation, or pigment-binder systems may be applied to the surface of the mineral and metallic fibers. Glass fibers can also be sized with a protein which then can be insolubilized and dyed with conventional protein dyes. Glass fibers are colored by coronizing, which involves preheating of the glass substrate to high temperatures to remove all organic materials followed by coloration with a pigment-binder system. The metallic fibers may also be colored through anodizing the metal (often aluminum) filament present or through pigmentation of the plastic layer coating the metal. The nature of the metal in the organometallic fibers determines their ultimate color.

18. Finishes and Finishing

Often fibers in textile substrates are deficient in one or more properties, or improved properties are desired for the substrate. Textile finishing provides a method whereby deficiencies in the textile can be corrected or specific properties can be introduced. Physical finishing techniques (dry finishing processes) or chemical finishing methods (wet finishing) are used. Physical finishing is usually carried out on the yarn or formed textile substrate, whereas chemical finishes can be added to the spinning bath prior to fiber formation for man-made fibers or applied to individual fibers, yarns, or completed textile structures.

PHYSICAL FINISHES AND FINISHING

Physical finishing methods for textiles include optical finishing, brushing and napping, softening, shearing, and compacting of the textile structure.

Optical Finishes

Luster may be imparted to a fabric by physical means. The techniques basically involve flattening or smoothing of the surface yarns using pressure. Beating of the fabric surface or passing the fabric between hard calendering rolls under pressure and with some friction will tend to flatten out the yarns and lower light scattering by the fabric surface, thereby improving reflectance and luster. Luster may be improved further if the calendering rolls are scribed with closely spaced lines which will be imprinted on the fabric to reinforce light striking and reflecting from the fiber surface. Similar techniques can be used to impart optical light interference patterns to the fabric (moiré). Thermoplastic fibers which can deform under heat and pressure can most readily be modified to impart luster.

Brushing and Napping

Physical delustering of a fabric as well as bulking and lofting of the fabric can be achieved by treatments which roughen the fiber surface or raise fibers to the surface.

Fiber raising processes such as brushing and napping involve use of wires or brushes which catch yarns in the textile structure and pull individual fibers partly from the yarn structure. The resulting fabric is warmer, more comfortable, and softer.

Softening and Shearing

During calendering or beating of a fabric interaction between individual fibers within yarns may be lessened and the textile structure softened. Also, when a smooth textile structure free of raised surface fibers or hairiness is desired, the fabric may be sheared by passing the fabric over sharp moving cutting blades or by passing the fabric over a series of small gas jets which singe and burn away raised fibers.

Compacting

During fabric formation processes, stresses often are introduced into a textile. Such stresses can be controlled by drying the finished fabric on a stenter frame, which controls the width of the fabric and the tension on the fabric during the drying process. A second method involves compression of the fabric structure, as in the Sanforizing process. In this process, the fabric and backing blanket (rubber or wool) is fed between a feed roller and a curved braking shoe, with the blanket being under some tension. The tension on the blanket is released after passing the fabric and blanket between the roller and braking shoe. The net result is compaction of the fabric being carried along in the system. Such a simple technique permits fabrication of the fabric of finished textile goods without fear of excessive shrinkage on laundering. Protein hair fibers such as wool, and thermoplastic fibers such as polyester, can be compacted by felting action. The scale structures on protein fibers entangle and stick on agitation, particularly in the presence of moisture. The resulting "ratcheting" effect causes the fibers to compact and felt. Many processes for wool take advantage of this effect, and nonwoven felt structures are produced by this method. Compaction of thermoplastic structure occurs when the fibers are raised to near their softening point. At a sufficiently high temperature the fibers shrink and contract, causing compaction of the textile structure.

CHEMICAL FINISHES AND FINISHING

Chemical finishes are chemical reagents or polymeric materials applied to textile structures by a number of methods. The major types of chemical finishes used on textile structures are listed in Table 18-1.

Table 18-1. Major Chemical Finishes.

Finishes Affecting Aesthetics, Comfort and Service	Protective Finishes
Optical finishes	Photoprotective agents and antioxidants
Hydrophilic and soil release finishes	
Softeners and abrasion resistant fibers	Oil and water repellants
Stiffening and weighting agents	Antistatic agents
Laminating agents	Biologically protective finishes
Crease resistant and stabilizing finishes	Flame retardants

Chemical finishes can be applied by a number of methods including padding (immersion in the treatment solution followed by squeezing to remove excess), spraying, printing, foam application, or vapor techniques. In addition, the finish can be added to the spinning bath prior to formation of man-made fibers. Of these methods, padding is most important. Many finished fabrics must be dried (to remove solvent) and cured (heated to cause a chemical reaction) before chemical finishing is complete. Thorough wetting of the fiber by the finish solution and spreading of the finish evenly over the fiber surface is critical in most cases to get the desired effect. The location of the finish on the surface or within the fiber is important, depending on the finish and its function.

Finishes Affecting Aesthetics, Comfort, and Service

Optical Finishes: Optical finishes do little to affect the color of a
textile substrate, but rather act to destroy or mask color centers. They
may either brighten the textile, making it more reflective, or deluster the
textile, making it less reflective, depending on the treatment.

Bleaches are usually chemical oxidizing or reducing agents that
brighten the textile by attacking unsaturated molecules that make the textile appear off-color. Chlorine bleaches such as sodium hypochlorite
(NaOCl) are strong oxidizing agents capable of destruction of color centers
on a textile substrate. Unfortunately, sodium hypochlorite is fairly non-
selective and attacks many dyes and finishes and certain fibers, causing
loss or change in color and a deterioration in fiber properties. Sodium
chlorite ($NaClO_2$) and peracetic acid (CH_3CO_3H) also are used as strong
oxidizing bleaches on some synthetic fibers at the mill to achieve desired
whiteness. Oxygen bleaches such as hydrogen peroxide (H_2O_2) and sodium
perborate (Na_3BO_3) are milder in oxidizing action and can be used on sensitive fibers such as wool. With hydrogen peroxide, pH adjustment is critical in getting the desired bleaching. Sodium perborate bleaching must be
carried out at elevated temperatures, although chemical activators may be
added to lower the effective bleaching temperature. Oxygen bleaches are
less likely than chlorine bleaches to damage the fiber and dyes present on
the fiber.

Reducing agents chemically reduce and saturate double bonds within
color centers and find limited use as bleaches in mill applications. Reagents such as sodium dithionate ($Na_2S_2O_6$) and sodium formaldehyde sul-
foxylate ($NaHSO_3 \cdot HCHO$) are commonly used as reducing bleaches and in compositions for stripping dyes from fibers. Fluorescent brightening agents are
colorless fluorescing dyes which mask yellow coloration on fabrics, and are
discussed in Chapter 17.

Delustering of fibers may be carried out through alteration of the
fiber surface or through addition of a light scattering and/or absorbing
agent to the fiber substrate. Chemical attack or etching of the fiber surface leads to a more irregular surface morphology, thereby increasing light
scattering and making the fiber dull. With man-made fibers, titanium
dioxide (TiO_2) is added to the spinning solution before fiber formation.
Titanium dioxide is an excellent light scattering agent, and this delusterant is effective when added at 0.5%-2.0% levels to the fiber. Titanium
dioxide can also be applied to the surface of natural fibers, if a binder

such as urea-formaldehyde resin is used to fix the delusterant to the fiber surface.

Hydrophilic and Soil Release Finishes: Hydrophilic (water-seeking) finishes that promote absorption or transport of water and aid in fiber wetting and soil removal in a textile construction are useful in many fiber applications. In man-made fibers surface-active agents such as nonionic polyethylene glycol derivatives can be added to the spinning solution prior to spinning, which will make the fiber more wettable and hydrophilic. Hydrophilic polymers such as copolymers containing acrylic acid can be fixed to fiber surfaces to provide improved water absorbency and to limit penetration of soils. Hydrolytic attack of the surface of hydrophobic man-made fibers such as polyester improves fiber wetting, moisture transport, and soil removal. None of the above absorbent techniques, however, are sufficiently effective to alter the absorbency properties of hydrophobic fibers like polyester enough to totally resemble naturally absorbent fibers like cotton and wool.

Softeners and Abrasion Resistant Finishes: Softeners and abrasion resistant finishes are added to textile structure (1) to improve aesthetics, (2) to correct for harshness and stiffness caused by other finishes on the textile substrate, and (3) to improve the ability of the fibers to resist abrasion and tearing forces. The softeners and abrasion resistant finishes are generally emulsions of oils or waxes, surface-active agents, or polymers that lubricate the surface of individual fibers in the textile substrate to reduce friction between fibers and permit them to pass over one another more readily. Emulsions of oils and waxes and related derivatives have been used in the past as softeners. Nonionic and cationic detergents act as softeners, but lack permanence. Emulsions of polyacrylates, polyethylene, or organosilicones impart softening properties and possess reasonable fastness. In addition, they may impart a more full hand to the textile.

Stiffening and Weighting Agents: Textile auxiliaries that stiffen and weight fabrics have included temporary and permanent sizes and metal salts applied alone or with a binding agent. The sizes stiffen the fabric through formation of bonds between fibers, particularly at fiber crossover points. Temporary sizes include starch, naturally derived gums, carboxymethylcellulose, and polyvinyl alcohol. Acrylic binders, polyvinyl chloride, and polyvinyl acetate emulsions act as permanent sizing agents. Metal salts may be used to weight certain fibers such as silk. If the metal salt has affinity for the fiber, as is the case for stannic chloride ($SnCl_4$) and silk, no binder is needed. Urea-formaldehyde or acrylic resins can act as suitable binders for fixation of nonpermanent metal salts onto the fiber surface. High intakes of metal salts used to weight textile structures may cause fibers to become brittle and to be sensitive to photochemically induced damage.

Laminating Agents: In recent years lamination of two textile structures together to form a composite structure has become very important. This technique requires use of adhesive materials that will not delaminate under normal use, including flexing and bending, shearing forces, and cleaning. The acrylics, polyurethanes, and a number of hot melt thermoplastic polymers are used as adhesives. Some adhesives have reactive groups which on curing lead to strong textile-adhesive covalent bonds in addition to the physical forces normally responsible for a good adhesive-fiber bond.

Crease Resistant and Stabilizing Finishes: When fibers are bent or deformed under various environmental conditions and then allowed to recover, the degree of recovery will depend on the morphology and inherent structure of the fiber. Most synthetic fibers show reasonable recovery from such deformation, whereas the cellulosics and, to a lesser extent, the protein fibers have poor recoveries, particularly under moist conditions. Crease resistant finishes for cotton and rayon have been developed that give much improved wrinkle recovery properties, whereas crease resistant finishes for wool are still under development, since the process is complicated by felting shrinkage.

Crease resistant finishes for cellulosics are chemical crosslinking agents and include wash-and-wear and durable press finishes. The wash-and-wear finishes are generally urea-formaldehyde or melamine-formaldehyde resins and are cured on cellulosics as flat goods in the mill, whereas durable press treatments consist of cyclic urea-aldehyde derivatives which are partially cured at the mill, after which curing is completed after the fabric is made into a garment in apparel manufacture. Other difunctional finishes, including epoxides, isocyanates, vinylsulfones, and aziridines, have been introduced as crease resistant finishes for cellulose. These have met with only limited success owing to their higher cost and other deficiencies. Since wool is already heavily crosslinked, a different approach to crease resistant finishing must be taken. The disulfide crosslinks in the textile structures must be chemically reduced followed by a setting treatment with bifunctional reagents such as capped diisocyanate derivatives. Chemical treatments for setting of thermoplastic man-made fibers are generally not used, since heat setting of these thermoplastic fabrics is an effective technique for imparting crease resistance. Urea-formaldehyde resins also are often used on cellulosics to impart dimensional stability to the textile structure and to prevent yarn slippage within the structure. Finishes capable of causing interfiber bonding can act as effective stabilizing finishes, although they may stiffen the overall structure.

Protective Finishes

   Photoprotective Agents and Antioxidants: Oxidative attack of fibers by oxygen in the air is particularly enhanced by the presence of moisture and light and leads to overall deterioration of the fiber, its properties, and useful lifetime. Weathering of natural cellulosic and protein fibers is well known, and many man-made fibers rapidly deteriorate unless a protective agent is added to the fiber. Protective agents are applied topically to natural fibers or added to the spinning solution prior to man-made fiber formation. They are basically of two types, (1) photoprotective agents that interact with light and (2) antioxidants that quench oxidative free radical attack. Photoprotective agents interact with ultraviolet and visible light absorbing and/or scattering the light and often releasing the absorbed energy harmlessly as heat. Photoprotective agents include metal oxides such as titanium dioxide and tin oxide, various metal salts, and ultraviolet absorbing organic compounds. Ortho-substituted benzotriazoles, triazines, and benzophenones are aromatic organic compounds that can harmlessly transfer the absorbed energy within the aromatic ring structure and thereby dissipate the energy eventually as molecular vibration and heat. Antioxidants interfere with oxidative free radical reactions formed through quenching and/or removal or the free radical from the fiber matrix so that it will not react further. Antioxidants are usually heavy metal salts that interfere with oxidation or aromatic organic compounds such as phenols and thiol derivatives which can readily donate a hydrogen to the radical, forming a stable free radical species that will not react further. Such antioxidants present in fibers are eventually exhausted and thereafter can provide no further protection.

   Oil and Water Repellents: Several classes of chemical agents exist that impart water and/or oil repellency when applied to textile substrates. Some finishes give water repellency only, whereas other finishes impart both water and oil repellency. Water repellent finishes are those which permit the fabric to continue to breathe after treatment, whereas water-proof treatments completely seal the spaces between individual yarns, as in the case of rubberized or unsaturated fatty acid-cured fabrics (oil cloths). Older water repellent treatments used derivatives of soaps and fatty acids to impart water repellency. Owing to their hydrocarbon nature they exhibit no oil repellency and actually are somewhat oleophilic (oil seeking). Soaps combined with zirconium, aluminum, or rare earth salts and methylol or pyridium salt derivatives of fatty acids were the major water repellent finishes used prior to World War II. With the development of organosilicon polymers after World War II, the methylated polysiloxanes were introduced as water repellent finishes. Also, fluorocarbon finishes generally are preformed organic polymers that have fluorine groups substituted on the side chains. It is important that the fluorine constituents be present at the end of the polymer side chains, with terminal trifluoromethyl (CF_3-) groupings being most effective. The unique properties of fluorine that enable fluorocarbon polymers to repel water, oil, and waterborne soils have contributed greatly to their extensive use on consumer goods under trade names such as Scotchgard and Zepel.

Antistats: Antistats or antistatic agents are finishes that can be applied to a fabric to aid in the dissipation of static charge buildup on the fibers. Antistats can be applied to the fiber as a temporary finish or added in the spinning bath prior to fiber formation to give a more permanent finish. Chemical crosslinking of an antistat applied to a textile structure will provide a permanent finish, also. Most natural fibers and regenerated natural fibers are hydrophilic and possess charged or polar groups on the fiber surface that can dissipate static charge to the atmosphere and prevent static buildup. Therefore antistat treatments are confined to the synthetic fibers such as nylon, polyester, etc. The antistats are surface-active agents related to detergents, ethylene oxide derivatives, silicones, or polar polymers such as polyamine resins. Because of their polar nature, they are able to bleed static charge from the fiber and dissipate it into the air.

Biologically Active Finishes: Natural fibers and regenerated natural fibers are subject to attack by various biological agents including bacteria, fungi, and insects. While synthetic fibers are not normally attacked by these biological agents, substrates of these fibers can act as a support for growth of bacteria and fungi on the fiber surface. Several metallic and organometallic salts, phenolic and anilide derivatives, and quaternary amine salts can inhibit growth of bacteria and fungi on fibers. Cellulosics most often require such treatment to prevent mildew and rot from feeding on and attacking the fiber substrate. Introduction of cyanide groups in cotton by graft polymerization of acrylonitrile on the substrate also prevents attack by rot. In the past, commercial insecticides such as DDT and Dieldrin were used to prevent insect attacks on cellulosic and protein fibers. Since many of the insecticides previously applied can no longer be used for such treatments, new insecticides which are not environmentally persistent or damaging have been developed and are related to the natural products, the pyretherins.

Flame Retardants: All textile fibers with the exception of glass are flammable. The degree of flammability is dependent on the chemical structure of the fiber, the construction of the textile substrate, and the environmental conditions present at the time of fiber ignition. A fiber is flame retardant when it self-extinguishes on removal of the flaming source. Certain fibers, including wool, modacrylic, aramid, and vinyon, are flame retardant by virtue of their inherent chemical structure and combustion characteristics. Thermoplastic synthetic fibers such as nylon and polyester are not self-extinguishing and continue to burn after ignition; however, owing to their melt-drip characteristics, the molten flaming polymer drops away from the fabric, causing the fabric to stop burning. Other fibers, such as the cellulosics, including cotton and rayon, burn readily and completely on ignition, leaving an ash which continues to oxidize and glow (afterglow) even after the flame is out. Blends of different types of fibers will show differing flammabilities which may not be directly related to an average flammability of the fibers in the blend. Flame retardants can be used on all fibers to improve their flammability characteristics further, with the most flammable showing the greatest degree of improvement on application of a flame retardant. Flame retardants for man-made fibers are generally introduced to the spinning solution prior to fiber formation, whereas natural fibers must be topically treated. Since the mechanism of flaming combustion of fibers varies with the fiber type, different retardants must be used for the various fiber types. The more flammable cellulosic fibers have received the greatest attention. Flame retardants can act in the gas phase or condensed phase of the burning fiber to interrupt oxidation and flaming and/or smoldering combustion. Organic compounds containing halogens (chlorine, bromine) and/or phosphorus, sometimes in conjunction with inorganic salts, have been found to be effective in many applications. Often different compounds in combination give a synergistic effect (an effect in excess of what might be expected from adding each component's contribution to flame retardation). Water-soluble inorganic salts containing phosphorus, boron, and ammonium have been found to be effective in "one-shot" nondurable applications, particularly on cellulosics.

FINISHES APPLIED TO FIBER CLASSES

Finishes for Cellulosics

Cellulosics have certain deficiencies in properties that require finishing to improve these properties and maximize consumer usage and acceptance. The most important cellulosic finishes include crease resistant and stabilizing finishes, soil release and softening finishes (in conjunction with crease resistant finishes), oil and water repellant finishes, biologically protective finishes, and flame retardant finishes.

Crease Resistant and Auxiliary Finishes: Cellulosics characteristically do not recover well from bending deformation, particularly in the presence of moisture, and crease resistant finishes have been developed for cellulosics to improve the wrinkle recovery of the cellulosic fabrics in the wet and/or dry states. Such finishes also stabilize cellulosics against relaxation shrinkage induced by mechanical forces during fabric formation. The crease resistant finishes used for cellulosics are for the most part derivatives of urea and various aldehydes. These resins chemically crosslink adjacent cellulose chains and provide a chemical memory within the cellulose which aids in recovery from bending deformation or wrinkling. The degree of dry and wet wrinkle recovery of the resin-treated cellulosics will differ depending on whether curing to achieve crosslinking was conducted in the dry state or in the swollen wet state.

Although crease resistant finishes for cellulosics were developed shortly after World War I, it was not until after World War II that they appeared extensively on consumer goods. The finishes introduced in the late 1940s were generally referred to as drip dry or wash-and-wear finishes. The resin-treated cellulosics were cured flat while they were still moist at the mill. This required that the treated fabric be permitted to drip dry to achieve the maximum desired recovery effect. Since the fabric was cured in the flat state before being made into a finished textile, the garment did not retain creases placed in the finished garment. Flat set items such as sheets and table cloths can be effectively made crease resistant by this process, however. In the early 1960s durable press finishes were introduced. After resin application and partial curing at the textile mill, the treated fabrics were sent to the apparel manufacturer. The apparel manufacturer fabricated the treated fabric into garments which were then fully cured in their finished state to give a textile product that retained its creases and recovered to its finished form after washing and tumble drying.

Although it is difficult to make a clear distinction between the resins used for the wash-and-wear and for durable press, certain generalizations can be made. Both types of resins are urea-aldehyde derivatives used in conjunction with possible other coreactants. The reactive functional groups in the resins are multiple N-methylol groups which undergo acid catalyzed reaction with hydroxyl groups in adjacent cellulose molecular chains to form bridging crosslinks to its unwrinkled state. Representatives of the major types of crease resistant resins for cellulosics are shown below:

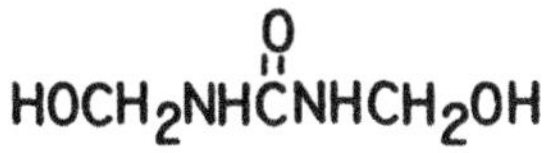

$$HOCH_2NHCNHCH_2OH$$
$$\overset{O}{\overset{\|}{}}$$

DIMETHYLOL UREA

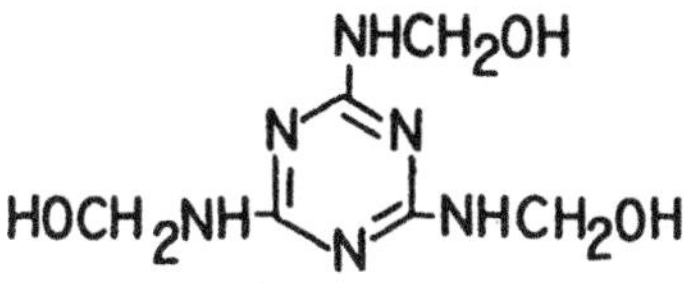

TRIMETHYLOL MELAMINE

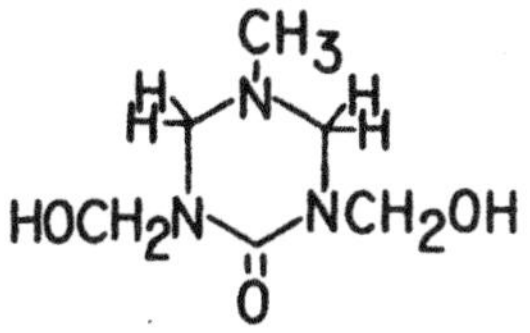

DIMETHYLOL-N-METHYL-TRIAZONE

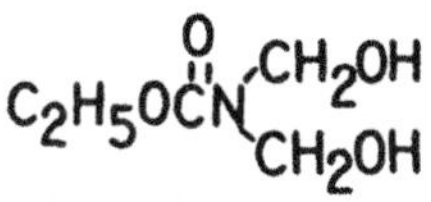

DIMETHYLOLETHYL-CARBAMATE

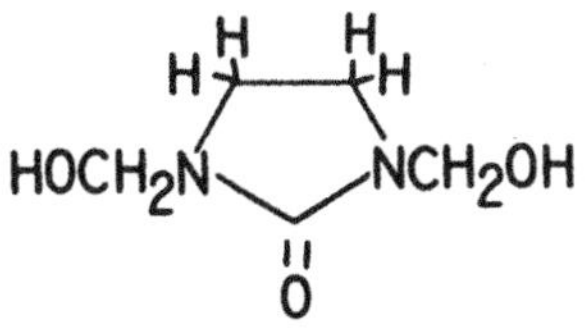

DIMETHYLOLETHYLENEUREA (DMEU)

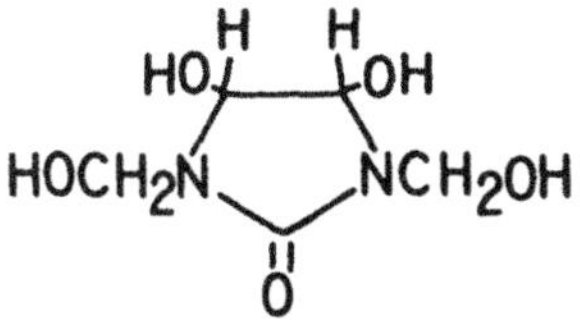

DIMETHYLOLDIHYDROXY-ETHYLENEUREA (DMDHEU)

In general, urea-formaldehyde condensates, N-methylolmelamines, and triazone-formaldehyde resins are used in wash-and-wear finishing. All suffer to varying degrees from chlorine retention and yellowing when in contact with sodium hypochlorite bleaches, which can cause damage to the cellulosic. In addition, the triazone resins give off a fishy odor under moist conditions. The ethyleneurea resins and the carbamates are used extensively in durable press treatments. In general, these resins are less susceptible to chlorine retention and undergo partial cure readily followed by a final cure. Latent or mild acid catalysts including inorganic and Lewis acid metal salts are added to the resin solution to achieve curing of all of these resins applied to cellulosics. A gaseous formaldehyde-sulfur dioxide vapor treatment for 100% cellulosics has been developed as well as a radiation curing treatment using N-methylolacrylamide. These resin treatments tend to lower the strength and overall abrasion resistance of the cellulosic fibers. Careful control of curing conditions and selection of catalysts as well as addition of softeners to the finishing solution can reduce this effect to some degree. Nevertheless, it has been necessary to blend thermoplastic heat settable fibers such as polyester with the treated cellulose to achieve acceptable wear characteristics. As a result poly- ester-cellulosic blends have made large inroads into markets where 100% cellulosic constructions were found previously. Soil retention on poly- ester-cellulosic blends treated with crease resistant resins has been a problem, and a series of polar hydrophilic soil release finishes have been developed to meet this need, as outlined before in this chapter. In recent years, low free formaldehyde durable press finishes have become important to protect workers from exposure to unreacted formaldehyde. In these formulations the methylol groups are capped with methyl groups, or a formaldehyde scavenger is added to the finishing solution.

Oil and Water Repellent Finishes: A wide range of oil and water re- pellent finishes are available for cellulosics and are described earlier in this chapter.

Biologically Protective Finishes: Cellulosics particularly used in outdoor conditions are susceptible to attack by microorganisms. Treatment of cellulosics with chlorinated phenolic derivatives, organometallics such as copper 8-hydroxyquinolate, cationic surfactants, or grafting with poly- acrylonitrile can reduce biological attack, although reductions in certain other properties may be expected.

Flame Retardant Finishes: Cellulosic fibers are among the more flam- mable fibers and are subject to rapid flaming combustion as well as slow

smoldering combustion. Flammability legislation and accompanying demand
for flame retardant cellulosics over the last 25 years has resulted in
development of a large number of flame retardant treatments for cellulosics, while only limited development of agents inhibiting smoldering combustion has taken place.

A number of temporary water-soluble flame retardant treatments for
cellulosics are used. For the most part, these treatments use mixtures of
inorganic salts, including borates, phosphates, silicates, and ammonium,
and aluminum, antimony, and other metal ion salts. Borax and boric acid
salt combinations are the most commonly used. Temporary flame retardants
find only limited use on "one shot" items that will not be laundered.
Permanent treatments are necessary for most cellulosic apparel and other
cellulosics which must be laundered. Permanent treatments for nonapparel
textile cellulosic substrates have involved antimony oxide in conjunction
with chlorinated hydrocarbons and crease resistant nitrogenous resins or
other resin binders. For apparel purposes, a number of resin treatments
based on tetrakis(hydroxymethyl)phosphonium chloride (THPC) or on THPC plus
sodium hydroxide (THPOH) have been used. THPC-amide based finishes include
THPC reacted with trimethylolamine and urea or THPC reacted with urea and
sodium hydrogen phosphate to form a complex polymer chemically fixed to the
cellulose. THPOH-amide finishes include THPOH-urea-trimethylolmelamine and
can be mixed with chlorinated hydrocarbons. THPC reacted in situ on the
cellulosic with gaseous ammonia followed by mild oxidation has become a
popular flame retardant finish in recent years. A dialkylphosphonamide,
Pyrovatex CP, can be effectively cured on cellulosics when used with a
triazine melamine resin to provide flame retardant properties. All of
these flame retardant treatments result in some loss to the wear characteristics and aesthetics.

Finishes for Cellulose Ester Fibers

A limited number of finishes are used to correct deficiencies in
acetate and triacetate. Delusterants, antioxidants, and/or antistat finishes as outlined previously are added to the spinning solution prior to
fiber formation. Plasticizers (nonvolatile aromatic esters) can also be
added to the spinning solution to improve the flexibility of the fiber.
When flame retardant treatment is required, haloalkylphosphonates are added
to the spinning solution or padded onto the yarns to effectively lower the
flammability of these fibers.

Finishes for Protein Fibers

Chemical Setting: Only limited finishing is carried out on protein fibers. Keratin fibers such as wool can be chemically set by chemical reduction of the disulfide crosslinks in the fiber, mechanically arranging the fiber in the shape (flat or creased) desired, and then reoxidizing and joining the cleaved disulfide bonds in the new configuration. Reducing agents commonly used are adducts of organic compounds with sodium dithionate ($Na_2S_2O_6$) or sodium bisulfite ($NaHSO_3$). The reduced disulfide (sulfhydryl) bonds are oxidized in time by oxygen in the air, but oxidation can be speeded up by application of a mild agent such as sodium bromate. This same process is used for permanent waving of human hair, which is also a keratin.

Shrinkproofing and Wrinkle Resistance Finishes: Wool and other keratin fabrics felt in the presence of moisture, agitation, and heat because of the surface scales on the fibers and resultant ratcheting action between individual fibers. Wool can be made resistant to felting shrinkage through modification of the scale structure by oxidizing the scales and/or by fixation of polymers on the scales. Oxidative treatments have included treatment by dry chlorination, wet chlorination, dichlorocyanuric acid, and permonosulfuric acid, as well as other oxidizing agents. The oxidizing agents attack the scales and partially destroy them, limiting their ability to cause felting. Although attack is mostly limited to the surface, significant damage and weakening occurs within the fiber. Polymer finishes that effectively can be spread and fixed on the fiber surface render wool shrink resistant even at uptakes as low as 1%-3%. Polymer is formed in situ on the fiber surface through reaction of step growth monomers such as diacid chlorides or diisocyanates with diamines (interfacial polymerization) or preformed polymers in solution or as an emulsion are applied to the wool and cured to chemically bond and fix the polymer to the wool. The presence of polymer on the fiber surface effectively prevents felting in a way that is not completely understood. The most satisfactory and effective treatments involve combined mild surface oxidation followed by polymer application. The usual shrinkproofing techniques have little effect on the wrinkle resistant characteristics of the wool. Finishes that contain chemically protected isocyanate groups have been developed that show some promise as durable press treatments for wool and have some ability to impart shrink resistant properties, too. Such treatments must be used in conjunction with chemical reduction of disulfide bonds in the fiber to set the fiber in the desired conformation in order to fully develop durable press properties.

Mothproofing Treatments: In the past, wool and other keratin fibers were protected from attack by moth larvae through use of mothballs (naphthalene or o-dichlorobenzene) or chlorinated insecticides. In recent years, the chlorinated insecticides used have been banned, and less toxic organic insecticides have been developed for use on wool.

Weighting Treatments: Silk is characteristically weighted through chemical treatment to give a heavier fabric. Silk is treated with metal salts such as tin chloride in acidic aqueous solution followed by washing with sodium phosphate and sodium silicate to form a complex tin phosphosilicate. High levels of salt will impart a harsh hand to the silk and further lower its resistance to sunlight-initiated oxidative attack.

Flame Retardant Treatments: Although the protein fibers are naturally flame retardant, additional flame retardancy is required in some applications. Wool can be effectively rendered flame retardant through treatment with halogenated organic acids, organophosphonates, and complex inorganic salts of zirconium, titanium, tungsten, etc.

Finishes for Polyamide Fibers

Although the polyamides are strong, tough fibers, they possess deficiencies with regard to oxidative attack by oxygen in the presence of light and/or heat which necessitates finishing to limit such attack. Static charge buildup on nylon is also a problem, and antistatic treatments are often applied to the fiber. Improved flame retardancy is desirable in some applications, and flame retardant treatments are often used. Selection of appropriate comonomers can introduce groups into polyamides which alter their dyeing characteristics.

Photoprotective Agents and Antioxidants: Several approaches have been used to stabilize the polyamides against oxidative attack in the presence of heat and/or light. Application or addition of antioxidant compounds such as stannic chloride or aluminum chloride plus base, copper and manganese salts, tungstates and molybdates, substituted phenols, catechols, organic phosphites, and combinations of these materials to the polymer melt or as finishes have been found to be effective in preventing such attack. Ultraviolet absorbers such as benzotriazoles and hydroxyphenones are effective in preventing light-induced oxidative attack. Titanium dioxide added to nylon as a delusterant will lower the absorption of light by the fiber but will also contribute to light-induced photodegradation through peroxide formation.

Antistatic Agents: Polar hydrophilic polymers applied as finishes or added to the polymer melt prior to fiber formation improve the static buildup characteristics of polyamides. Polyethylene glycols and acid-containing vinyl and acrylic copolymers have been used, as have insoluble salts of amphoteric and cationic detergents.

Flame Retardant Finishes: Polyamide fibers can be most effectively flame retarded by thiourea-aldehyde condensate treatments in presence and absence of durable press finishes. Amino phosphorazines, ammonium bromide-binder combinations, and antimony oxide with halogen-containing polymers have also been found to be effective. Tris(2,3-dibromoprophyl)phosphate (better known as Tris) is effective as a halogenated organophosphorous compound flame retardant, but due to its mutagenic and possible carcino-genic characteristics it has been excluded from use.

Finishes for Polyester Fibers

The major finishes for polyester fibers include photoprotective agents and antioxidants, antistatic agents, soil release finishes, antipilling finishes, and flame retardant finishes. Many of these chemicals can be added to the polymer melt prior to spinning. The finishes can also be added topically during dyeing and finishing.

Photoprotective Finishes and Antioxidants: Compounds that are used to stabilize polyesters against heat- and/or light-induced oxidative attack are similar to those used for nylon and other synthetics. They include metal salts as well as benzotrizoles, benophenones, and hindered phenols.

Antistatic Finishes: Owing to its hydrophobicity, polyester builds up static charge readily. Finishes have been developed for polyester that in-crease the hydrophilicity and ionic character of the fiber and permit more ready dissipation of static charge from the fiber surface. These treat-ments include lauryl phosphate, morpholine, various polyethylene glycols, organosilicones, and polyamine resins.

Soil Release Finishes: Owing to its combined hydrophobic and oleo-philic character, polyester is soiled readily by oily soils. If these soils penetrate into the fiber, they are difficult to remove by laundering. In order to improve the removal of oily soils from polyester, oil repelling soil release finishes have been developed to limit the penetration of oily soils into the fiber, thereby making their removal easier during launder-ing. Perfluoroalkylamides and other fluorocarbon derivatives have been used for this purpose, but more inexpensive hydrophilic finishes such as copolymers of acrylic acid, polyhydroxy compounds, and carboxymethylcellulose have been used more extensively. Since polyester is often used with cellulosics in blends in which the cellulosic must be treated with a durable press finish, combined durable press-soil release finishes have been prepared and are useful and very effective on such blends.

Antipilling Finishes: Because of its strength, polyester is subject to pill formation during wear. To make polyester more pill resistant, weaker and therefore less pillable, polyester fibers have been prepared by using lower molecular weight polymers or through partial hydrolysis of the polyester after textile substrate formation.

Flame Retardant Finishes: With the implementation of various flammability regulations in the U.S., improved flame retardant properties for polyester were required particularly with regard to the continued burning of melted polyester falling from burning fabric. In order to impart self-extinguishing characteristics to the polyester, two approaches have been used--introduction of halogen- and/or phosphorus-containing comonomers to the polymer structure or introduction of a halogenated phosphonate or related compounds to the melt prior to spinning. In both cases, antimony oxide may be added to improve the overall flame retardant characteristics.

Tris(2,3-dibromopropyl)phosphate (better known as Tris) was an ideal flame retardant for polyester and was used extensively for this purpose. Unfortunately, Tris was found to be mutagenic and a potential cancer-producing agent and had to be removed from the marketplace. As a result of the ensuing confusion, all topical flame retardant finishes have become suspect, and certain flammability requirements have been relaxed to allow polyester to pass the flammability test for children's sleepwear without further treatment.

Other Finishes: Comfort and the moisture related properties of polyester are of great interest, and finishes and treatments that improve these characteristics should be of great importance in coming years. Polyester fibers can be rendered more water wettable by surface treatment with basic solutions or through application of a hydrophilic finish to the polyester surface. Such finishes improve the water wicking and water transport characteristics of the polyester. Base hydrolysis has the added benefit of giving the polyester a crisp silk-like hand. Fine denier base-treated polyesters are produced in Japan that are effectively used as a silk substitute.

Finishes for Acrylic Fibers

Antistatic treatments for acrylic fibers include polyglycol esters, fatty acids, and hydroxyl and amino derivatives that can be crosslinked with aldehydes or epoxy groups. Certain of these antistats as well as cationic surfactants can be used as softening agents to impart a softer hand to acrylic fibers. Modacrylics are flame retardant and self-extinguishing and usually do not require an additional flame retardant treatment. Acrylic fibers can be made flame retardant by use of a halogenated comonomer with acrylonitrile or through addition of an organic phosphorus derivative or halogenated material to the polymer solution prior to spinning.

Finishes for Polyolefin Fibers

Antistatic finishes and antioxidants normally used on synthetic fibers are added to the polymer melt prior to polyolefin fiber formation. The most common antioxidants used include hindered substituted phenols, organometallic antioxidants, and substituted phenols. Blending of comonomers with polyolefin prior to fiber formation is used not only to improve dyeability but also to plasticize the fiber and improve other fiber properties. Polyolefins may be effectively made flame retardant through incorporation of a metal oxide such as antimony oxide in conjunction with a brominated hydrocarbon or brominated organophosphate.

Finishes for Vinyl Fibers

Vinyl fibers except for vinal and vinal-vinyon do not generally undergo specialized finishing, although antioxidants and antistatic agents may be used. Vinal and vinal-vinyon matrix fibers can be treated with urea-aldehyde based durable press finishes to improve their wrinkle recovery, and vinal can be made flame retardant with phosphorus-containing finishes used for cellulosics.

Finishes for Elastomeric Fibers

The elastomeric fibers often have antioxidants and antistatic agents of the types outlined previously incorporated into them to protect the fiber against oxidative attack and to reduce static buildup, respectively.

Finishes for Mineral and Metallic Fibers

A limited number of finishes are used on mineral and metallic fibers. Glass fibers are often surface treated with agents such as chromium salts or silanes to improve their ability to adhere to organic adhesive materials used in polymer-glass fiber composites. Various organic starches, proteins, and synthetic polymer sizings are applied to glass fabrics to protect the individual fibers and to lower abrasion between individual fibers.

V. Textile Maintenance

19. Textile Soiling and Soil Removal

The maintenance of a textile product after purchase is of prime interest to the consumer and to commercial fabric care operations. The major factor to be considered is cleaning and soil removal of the textile during continued use. In order to have a fuller understanding of the cleaning process, one must examine the nature of textile soils, detergency, and soil removal, and the wet (laundering) or dry (solvent) cleaning processes used.

TEXTILE SOILS

Soils come from a number of sources in the environment that textile structures are subjected to during wear and use. These soils include (1) solid particulate matter (clays, minerals, soot), (2) oil-borne soils (fats, greases, etc.), and (3) water-borne soils (water-soluble salts, etc.). Solid particulate matter such as clays, metal oxides, and soot is often mixed with water- and oil-borne soils but can also soil a textile alone through application in the dry state. When applied from the dry state, these solid soils can often be removed by mechanical action such as brushing and shaking.

Clays in general are complex inorganic silicates with color derived from the structure of the silicate. Oil-borne soils are organic hydrocarbons or related derivatives which are soluble in oils. The aliphatic and aromatic hydrocarbons and fatty acid esters of glycerol are the most predominant oil-borne soils. Less polar hydrocarbons such as mineral oil are more easily removed from textile substrates than are the more polar glycerol esters of fatty acids. Carbon-based matter such as soot is not completely soluble in most hydrocarbon solvents or oils but must be considered an oil-borne soil. These soils can be removed by solvent (dry) cleaning or through emulsification and removal in laundering systems. Water-borne soils are usually water-soluble inorganic and organic salts or natural proteins and starches and can be readily removed by water-based laundering systems. Soils from foods can be oil- and/or water-borne soils, depending on the composition of the particular food.

DETERGENCY AND SURFACTANTS

Detergency and Soil Removal

Detergency is a term used to specify the ability of an agent to lift and remove soil from a substrate and to suspend the soil within the cleaning media. Agents which aid directly in soil removal are called detergents or surface active agents (surfactants). Since the term detergent has come to mean complex laundry formulates containing several components, the soil lifting components of such formulations will be referred to as surfactants here to avoid confusion.

Surfactants are compounds containing an oleophilic (oil-searching) hydrocarbon tail and a hydrophilic (water-seeking) polar head that can effectively aid in wetting of a soiled textile surface, in penetration and removal of the soil from the surface, and in suspension of the soil in the liquid medium. Surfactants are materials which effectively make the transition between the relatively nonpolar hydrocarbon soil and the polar cleaning medium such as water. The hydrocarbon tail of the surfactant associates with the surface of the oily soil, whereas the polar head of the surfactant associates with the aqueous medium, thus making a transition from the oily soil to the aqueous media. When oily soil is lifted from a fiber by the surfactant, the oily soil-detergent combination is suspended as small particles in the medium through micelle formation.

The low-energy micelle formed must be sufficiently stable to permit its removal in the laundering process. Mineral soils, being partially hydrophilic in nature, undergo a more complex process in soil removal. The soil mixes with the surfactant to form a liquid crystal. Additional surfactant forms a complex micelle which includes myelinic tubes to provide sufficient surface area to remove and stabilize the solubilized soil.

Surfactants

Surfactants are divided into five major classes: soaps, anionic, non-ionic, cationic, and amphoteric surfactants. Each contains a hydrocarbon tail and a polar head. Typical surfactants of each class are represented below:

SOAP $R-COO^- Na^+$

ANIONIC $ROSO_3^- Na^+$ $SO_3^- Na^+$

NONIONIC $RCOO(CH_2CH_2O)_n H$ $O(CH_2CH_2O)_n H$

CATIONIC

AMPHOTERIC $R\overset{H^+}{\underset{H}{N}}CH_2CH_2COO^-$

R = LINEAR $C_{12} - C_{18}$ HYDROCARBON CHAIN

Soap: Soap has been known since antiquity as a surfactant for removal
of soil from textiles. Soap is readily made by basic hydrolysis (saponification) of animal fats (fatty esters of glycerol). Soap is the resultant
sodium salt of the fatty acids, with the composition depending on the
source of fatty acid esters. Soap suffers from one major deficiency as a
surfactant: in hard water containing calcium and magnesium cations, the
sodium ion in soap is replaced by these multivalent ions to form insoluble
salts which cannot act effectively as surfactants.

Anionic Surfactants: Anionic surfactants by definition contain an
anion (negative ion) as the hydrophilic head of the detergent and are
usually sodium, potassium, or ammonium salts of organic sulfonates or sulfates such as alkylbenzene sulfonates or alkyl sulfates. Anionic surfactants are effective in removal and suspension of oily soil and remain
soluble in the presence of calcium and magnesium ions. For this reason,
they are preferred over soap and are the most used surfactant in laundry
formulations. The alkylbenzene sulfonates and particularly sodium dodecyl-
benzene sulfonate are used in such formulations. In the 1950s, foaming
problems in water supplies were attributed to these surfactants due to
their low degree of biodegradability. Studies at that time showed that
branching of the alkyl group substituted on the benzene ring was responsible for this problem. Reaction conditions for formulation of these
surfactants were changed so that the more biodegradable linear alkyl
derivative was produced, thereby correcting the problem. Since anionic
surfactants tend to foam readily, they are seldom used textile processing.

Nonionic Surfactants: The nonionic surfactants contain a polar head
which provides sufficient hydrophilicity to give detergent activity. Polymers of ethylene oxide (called polyethylene glycols or polyethoxyethanols)
commonly are used as the polar head attached as the the alkylbenzene or
alkyl moiety to form the nonionic surfactant. The hydrophilicity of the
ethoxy repeating unit comes from the hydrogen bonding capability of the
ether oxygen with water. The nonionic surfactants are used in conjunction
with anionic surfactants in some laundry formulations and as wetting agents
in many textile dyeing and finishing wet processes.

Cationic Surfactants: Cationic surfactants possess a positive cation
and are usually quaternary amine salts. Owing to their high cost, they are
less important than anionic and nonionic surfactants in detergent formulations. They are mainly used as fiber wetting agents and as bacteriostats
and fabric softeners in selected applications.

Amphoteric Surfactants: Surfactants that have both positively and negatively charged hydrophilic groups within the molecule are referred to as amphoteric surfactants. The detergency of these surfactants varies with pH, and they show bacteriostatic activity at appropriate pH. Amphoteric surfactants are effective leveling agents and aid in controlled diffusion of dyes and finishes onto the fiber.

LAUNDERING AND LAUNDRY FORMULATIONS

Laundering

Laundering is essentially a wet cleaning process in water solvent in the presence of a detergent formulation. The physical parameters, agitation and temperature affect the ease and effectiveness of soil removal. Agitation permits the aqueous surfactant solution to flow through the textile structure, conveying the surfactant to the soil, and aids in removal of emulsified soil from the fabric. As the laundering temperature increases, the surface activity of the surfactant solution increases, which in turn increases the ease and rate of soil removal. Although higher temperatures markedly improve soil removal, the maximum temperature that can be used may be tempered by a number of factors, including the stability of the textile and its washfastness.

The nature of the impurities in the water has a major effect on soil removal from a textile. If the water is hard and contains significant amounts of calcium and magnesium salts as carbonates, sulfates, or chlorides, these salt ions will interfere with the soil lifting action of the surfactant unless appropriate water softening agents are added. Dissolved iron salts or the presence of clays, silts, and other colorants can interfere with cleaning, also.

Laundry Formulation

Laundry powder formulations or synthetic detergents (often called syndets) are complex mixtures of surfactant and other materials including many of the following (average range of composition in syndets in parentheses): surfactant (10-30%), builders and chelating agents (5%-40%), anti-soil-redeposition agents (0.5%-2%), corrosion inhibitors (5%-10%), foam stabilizers and antifoaming agents (0%-5%), electrolytes and fillers (5%-40%), oxygen bleaches (0%-25%), fluorescent brighteners and colorants (0.1%-1%), bacteriostats (0%-2%), perfumes (0%-1%), and moisture (0-10%). The composition of the detergent formulation will change with the manufacturer and intended use. The liquid detergents are aqueous solutions of similar composition to detergent powders with the following exceptions: (1) the anionic surfactants present will tend to be the more soluble miscible potassium, ammonium, or alcohol amine salts; (2) the nature of added foam stabilizers will differ, and (3) the amount of builder present will be lower. A series of specialized product formulations, including enzymes, bleaches and brighteners, water softeners, etc., also are on the market as auxiliary cleaning agents.

Builders: Builders are salts added to a detergent composition to improve the effectiveness of the surfactant present through complexation or precipitation of calcium and magnesium and other multivalent salts. The builders act through complexation (chelation) with these cations to form a stable complex or through reaction with the cations to form an insoluble salt that precipitates from the wash bath. Complexing builders include the sodium polyphosphates (trisodium tripolyphosphate and tetrasodium pyrophosphate), amine carboxylates, citrates, carboxylate polymers, and zeolite ion exchange resins. These builders all complex with calcium and magnesium ions to form water-soluble complexes (chelates) or suspensions that do not interfere with the action of the surfactant. The polyphosphate builders are the most effective builders but have come under increasing pressure over the last decade due to their role as biological nutrients and contributors to algae growth. The percentage of polyphosphates used in detergent formulations has declined in recent years, but substitutes that are as effective have been difficult to find at a comparable cost. Nitrilotriacetic acid was introduced in the late 1960s as a builder, but adverse factors, including its possible activity as a carcinogen, caused it to be withdrawn from the market.

Precipitating builders include sodium bicarbonate and sodium carbonate, sodium sesquicarbonate (a mixture of the two), and the borate salts. These builders provide basicity and react with calcium and magnesium ions to form the insoluble carbonates or borates. These builders are not as effective as chelating builders. After repeated washes they leave deposits of carbonates mixed with soil on the textile being cleaned. They also may decrease the water absorbency of the textile with time.

Anti-Soil-Redeposition Agents: Soil removal is a dynamic process in which suspended soil may be redeposited on the textile as well as removed during the laundering process. Addition of agents with appropriate soil-repelling functional groups inhibits such redeposition. Carboxymethylcellulose is an inexpensive negatively charged water-miscible polymer that forms a thin deposit or coating on the textile and repels the charged soil-detergent micelle. Other polar or charged water-miscible polymers such as polyvinylpyrollidone are particularly useful on synthetics as effective anti-soil-redeposition agents and can be incorporated with carboxymethyl-cellulose to improve the overall effectiveness of anti-soil-redeposition particularly on synthetic-natural fiber blends.

Corrosion Inhibitors: The basicity and reactivity of ingredients found in laundry formulations lead to attack and corrosion of various metal parts in laundry equipment. To minimize this effect, the sodium silicates are added to the detergent formulation.

Foam Modifiers: Excess foaming during laundering can occur readily due to agitation and can lower the overall effectiveness of soil removal. On the other hand, the consumer views moderate and stable foam formation during laundering as an indication of detergency and soil removal. Two approaches have been used to provide products which meet both of these concerns. Antifoaming agents such as long-chain aliphatic alcohols, emulsified terpenes (naturally occurring alcohols), and organosilicones are used in conjunction with foaming surfactants to lower and moderate foam formation. The second approach has been the use of detergentlike derivatives that modify and stabilize foaming in conjunction with surfactant. These foam modifiers include monoalkylolamine adducts of fatty acids and their polyethylene oxide derivatives.

Electrolytes and Fillers: Inorganic salts such as sodium sulfate are added to laundry formations to bring them up to uniform cleaning strength and to provide appropriate measurable quantities for addition in laundering by the consumer. These materials may be considered fillers but also are electrolytes in solution that serve to enhance to some degree the migration and action of the surfactant as well as improve the physical character-istics of the product.

Bleaches and Fluorescent Brighteners: Oxygen bleaches such as sodium perborate are often added to enhance the whitening power of the formulation through destruction of color centers remaining on the fabric. Fluorescent brighteners are added to nearly all synthetic fibers in manufacture to cover yellow coloration through blue fluorescence of these colorless dyes in the light. Fluorescent brighteners added to laundry formulations are mixtures of brighteners which have affinity for all fiber types commonly found in a wash load. Bleaches and brighteners also can be purchased and used separately to enhance whitening of the textile substrate. Additional information concerning the structure and action of bleaches and brighteners appears in Chapter 18.

Germicides: Biologically active germicides are added to some syndets and are particularly important in low-temperature laundering, where biological agents are not destroyed by heat. The germicides include cationic surfactants and phenol derivatives as well as natural products such as contained in pine oil. Chlorine bleaches also act as germicides in laundering.

Perfume: Perfumes are added to laundry formulations to mask odors of other ingredients and to convey a pleasant odor which may be suggestive of a natural fragrance or of a clean wash. The perfume has essentially nothing to do with effective soil removal but adds to product aesthetics and aids in consumer acceptance of the product.

Fabric Softener: Fabric softeners are product compositions containing a cationic or nonionic surfactant or alkoxyalkylamide, and they may be applied during the laundering rinse cycle or transferred to the textile during drying from an inert cellulosic or polyurethane substrate. Liquid softener compositions usually contain alcohol (0%-2%) (to increase solubility of the softeners), softeners (2-8%), surfactant (0-2%), electrolytes (0%-0.25%), fluorescent brighteners (0%-0.32%), germicides (0%-2%), colorants (0%-0.2%), and perfumes (0%-2%). The rest is water. Fabric softener components used in dryers are less complex, containing softener with a carrier and perfume in an inert substrate.

Starches: Starches used to give a textile stiffness and body are usually added during the rinse or applied as a spray after washing. Starches include naturally derived starch, starch derivatives, and acrylic polymer emulsions.

Enzymes: In the late 1960s enzyme presoaks and laundry products containing enzymes were introduced. The proteolytic enzymes contained within the products must have a presoak period to be effective. They act as catalysts is speeding the hydrolytic attack of protein and carbohydrate components in soils, breaking them down into more easily removed decomposition products. Since oily soils are not readily attacked by the enzymes, their use and effectiveness is limited.

DRYCLEANING

Drycleaning is not carried out under dry conditions at all but rather uses a solvent other than water in the cleaning method. The cleaning is carried out in petroleum hydrocarbon (Stoddard solvent), in a chlorinated solvent (tetrachloroethylene and trichloroethylene), or in a fluorohalocarbon (Freon). Although the solvents will effectively remove saturated or oily soils, the solvent is charged with water plus surfactant to aid in soil emulsification of the more hydrophilic soils. Tetrachloroethylene is the predominant solvent used in drycleaning in the United States. The drycleaning process involves prespotting by an appropriate method to clean any badly soiled areas on the textile. The textile is immersed in the cleaning fluid, and the fluid is circulated through the textile and then filtered through activated charcoal and diatomaceous earth to remove impurities. After a certain period of continued use, the drycleaning solvent is redistilled to remove residual oils and so forth and then recharged with water and detergent. Drycleaning solvents cause less fiber swelling and deformation and have less tendency to remove dye from the textile. Less agitation is involved than in laundering, and drycleaning is therefore preferred for textiles in which water-induced dimensional change will occur. Care must be taken in drycleaning some textiles, since damage may occur due to attack of the fiber, of finishes on the fiber, or of one or more components in the fiber structure.

Appendix: Suggested Further Reading

FIBER THEORY, FORMATION, AND CHARACTERIZATION and FIBER PROPERTIES

R. S. Asquith, Chemistry of Natural Protein Fibers, John Wiley, N.Y., 1977.

C. B. Chapman, Fibres, Butterworth, London, 1974.

C. A. Farnfield and D. R. Perry, Identification of Textile Materials, The Textile Institute, Manchester, 1975.

E. P. G. Gohl and L. D. Vilensky, Textile Science, Longman Cheshire, Melbourne, 1980.

M. Grayson, Encyclopedia of Textiles, Fibers, and Nonwoven Fabrics, J. Wiley, N.Y., 1984.

A. J. Hall, The Standard Handbook of Textiles, 8th Ed., Newnes-Butterworth, London, 1975.

A. F. Happey, Applied Fibre Science, Vol. I-III, Academic Press, N.Y., 1978-79.

M. Lewin and J. Peston, Handbook of Fiber Science and Technology, Vol. III, Marcel Dekker, N.Y., 1984.

H. F. Mark, S. M. Atlas and E. Cernia, Man-made Fibers: Science and Technology, Vol. I-III, Interscience, N.Y., 1967-1968.

R. W. Moncrieff, Man-made Fibres, 6th Ed., Heywood, London, 1975.

B. F. Smith and I. Block, Textiles in Perspective, Prentice-Hall, N.J., 1982.

M. A. Taylor, Technology of Textile Properties: An Introduction, 2nd Ed., Forbes Publications, London, 1981.

YARN AND TEXTILE SUBSTRATE FORMATION

B. C. Goswami, J. G. Martindale, and F. L. Scardino, Textile Yarns: Technology, Structure, and Application, John Wiley, N.Y., 1977.

F. Happey, Contemporary Textile Engineering, Academic Press, N.Y., 1982.

T. Ishida, Modern Weaving, Theory and Practice, Osaka Senken Ltd., Tokyo, 1979.

G. Lubin, Ed., Handbook of Composites, Van Nostrand Reinhold Co., N.Y., 1982.

P. R. Lord, The Economics, Science and Technology of Yarn Production, 2nd Ed., The Textile Institute, Manchester, 1981.

P. R. Lord and M. H. Mohamed, Weaving: Conversion of Yarn to Fabric, Merrow, Watford, 2nd Ed., 1982.

R. Marks and A. T. C. Robinson, Principles of Weaving, The Textile Institute, Manchester, 1976.

P. Schwartz, T. Rhodes, and M. Mohamed, Fabric Forming Systems, Noyes Publications, Park Ridge, N.J., 1982.

J. A. Smirfitt, An Introduction to Weft Knitting, Merrow, Watford, 1975.

D. G. B. Thomas, An Introduction to Warp Knitting, Merrow, Watford, 1976.

J. J. Vincent, Shuttleless Looms, The Textile Institute, Manchester, 1980.

D. T. Ward, Ed., Modern Nonwovens Technology, Texpress, Manchester, 1977.

PREPARATION, DYEING AND FINISHING PROCESSES and TEXTILE MAINTENANCE

F. W. Billmeyer and M. Saltzman, Principles of Color Technology, 2nd Ed., J. Wiley & Sons, N.Y., 1981.

C. L. Bird and W. S. Boston, Eds., Theory of Coloration of Textiles, Dyers Company Publication Trust, Bradford, 1975.

A. Datyner, Surfactants in Textile Processing, Marcel Dekker, N.Y., 1983.

C. Duckworth, Ed., Engineering in Textile Coloration, Dyers Company Publication Trust, Bradford, 1983.

M. Lewin and S. B. Sello, Eds., Handbook of Fiber Science and Technology, Vol. I and II, Marcel Dekker, N.Y., 1983-84.

L. W. C. Miles, Ed., Textile Printing, Dyers Company Publication Trust, Bradford, 1981.

D. M. Nunn, Ed., The Dyeing of Synthetic-Polymer and Acetate Fibers, Dyers Company Publication Trust, Bradford, 1979.

F. Sadov, M. Korchagin, and A. Matetsky, Chemical Technology of Fibrous Materials, MIR Publishers, Moscow, 1978.

E. R. Trotman, Dyeing and Chemical Technology of Textile Fibers, 6th Ed., Griffin, London, 1984.

Index

TEXTILE IDENTIFICATION, CONSERVATION, AND PRESERVATION

by

Rosalie Rosso King

Western Washington University

Textile identification, preservation and conservation methods, as presented in this book, include both simple and complex processes that relate to techniques used in both the arts and sciences. The process of identification and conservation must be exacting to insure that mistakes are not made which would impair the beauty or lasting quality of the textile piece. An in-depth understanding of fibers, assemblages of fibers, and fabrics, plus a study of their reaction to cleaning procedures, is necessary to insure that future generations will have the opportunity to enjoy textile pieces currently being produced as well as those collected from the past.

Because each culture has developed textile artifacts, the world is rich in a diversity of pieces and an abundance of soft goods. Yet some items were made in very limited quantities for extremely specialized end uses. Some have required years of often painful labor for their construction. The primary aim in the writing of this book is to provide identification, cleaning and conservation techniques fór all textile fibers and fabrics found today. A review of textile and related research findings through the early 1980s has been included, as the importance of clean, well-cared-for textiles cannot be overlooked in relation to the management of a collection, museum storage, or to the expected life span of actual pieces. However, some textile pieces containing historically important soil or foreign matter, or those that would be altered or changed, should be left untouched. These cases are also discussed.

Because of their increased use in recent years, man-made fibers and fabrics are given detailed coverage in the book. The condensed table of contents given below lists **chapter titles and selected subtitles**.

ISBN 0-8155-1033-0 (1985)

356 pages

CORROSION RESISTANT MATERIALS HANDBOOK
Fourth Edition

Edited by

D.J. De Renzo

The Fourth Edition of the *Corrosion Resistant Materials Handbook* has been completely revised and vastly expanded, based on the latest available technical data. This well-established and successful reference volume, first published in 1966, will provide useful information which will enable the concerned engineer or manager to cut losses due to corrosion by choosing suitable *commercially available* corrosion resistant materials for a particular application.

The great value of this outstanding reference work lies in the extensive cross-indexing of thousands of substances. The more than 160 tables in the book are arranged by types of **corrosion resistant materials.** The **Corrosive Material Index** is organized by *corrosive chemicals* and other *corrosive substances.* A separate Trade Name Index and a Company Name and Address Listing are also included.

The various sections in the book cover selected categories of corrosion resistant materials, such as synthetic resins and polymers; rubbers and elastomers; cements, mortars, and asphalts; ferrous alloys; nonferrous metals and alloys; and glass, ceramics, and carbon-graphite. A separate section contains 13 tables which compare the anticorrosive merits of a cross section of commercial engineering and construction materials essential to industry. The tables in the book represent selections from manufacturers' literature made at no cost to, nor influence from, the makers or distributors of these materials.

A condensed table of contents listing **chapter titles and selected subtitles** is given below. Parenthetic numbers indicate number of tables per topic.

ISBN 0-8155-1023-3 (1985)   8½" x 11"   962 pages

TEXTILE WET PROCESSES
Vol. 1
Preparation of Fibers and Fabrics

by

Edward S. Olson
Clemson University

Textile wet processing, as it applies to the preparation of fibers and fabrics, is described. This is the first of a three-volume series which will cover the main areas of textile wet processing—preparation, coloration, and finishing.

Basically, textile wet processes should be considered a chemical processing industry. Recognizing and understanding the interactions of many inorganic and organic chemicals with simple and complex polymers is a critical aspect of this innovative and dynamic industry. It is, thus, important that those practicing the art or considering entering the market have a thorough background in the fundamentals of the chemistry involved.

A sequential approach has been taken to the processes described. The scientific basis is explained followed by a description of current practice and associated equipment. Natural as well as synthetic fibers and fabrics are discussed, as are water handling and economy, heat-setting, singeing, desizing, scouring, bleaching, and calculations useful for wet processing.

The book is well illustrated with photographs, schematic diagrams, figures, and tables. A condensed table of contents is listed below.

ISBN 0-8155-0939-1(1983)

205 pages

FABRIC FORMING SYSTEMS

by

Peter Schwartz
Cornell University

Trevor Rhodes
Levi Strauss & Co.

Mansour Mohamed
North Carolina State University

This basic technical text on fabrics and fabric formation highlights both woven and knitted fabrics, as well as the major categories of nonconventional fabrics. The main fabric forming systems, and the structures produced, in weaving, knitting, and nonconventional fabric production are detailed.

The book covers the preparation of yarn for weaving and knitting, loom mechanisms, classification of looms, woven fabric design, weft knitting and weft knit structures, warp knitting and warp knit structures. Included also is a brief introduction to nonwoven fabric web formation and different bonding or manufacturing systems such as saturation bonding, print bonding, spunbonding, needle punching and stitch bonding. Tufting and flocking are also discussed.

The book will be particularly suitable for textile managers, students, or professionals of nontextile background seeking an overview of textile manufacturing.

Detailed diagrams illustrate the written text throughout the book. A table of contents listing **chapter titles and selected subtitles** is given below.

ISBN 0-8155-0908-1 (1982)

175 pages

Printed and bound by CPI Group (UK) Ltd, Croydon, CR0 4YY

03/10/2024

01040430-0003